ISBN 978-1-333-34177-0
PIBN 10492797

1 MONTH OF
FREE
READING

at

www.ForgottenBooks.com

By purchasing this book you are
eligible for one month membership to
ForgottenBooks.com, giving you
unlimited access to our entire
collection of over 1,000,000 titles via
our web site and mobile apps.

To claim your free month visit:

www.forgottenbooks.com/free492797

English
Français
Deutsche
Italiano
Español
Português

www.forgottenbooks.com

Mythology Photography **Fiction**
Fishing Christianity **Art** Cooking
Essays Buddhism Freemasonry
Medicine **Biology** Music **Ancient
Egypt** Evolution Carpentry Physics
Dance Geology **Mathematics** Fitness
Shakespeare **Folklore** Yoga Marketing
Confidence Immortality Biographies
Poetry **Psychology** Witchcraft
Electronics Chemistry History **Law**
Accounting **Philosophy** Anthropology
Alchemy Drama Quantum Mechanics
Atheism Sexual Health **Ancient History**
Entrepreneurship Languages Sport
Paleontology Needlework Islam
Metaphysics Investment Archaeology
Parenting Statistics Criminology
Motivational

CARTWRIGHT LECTURES

1894

ON

DIGESTIVE PROTEOLYSIS

BEING THE CARTWRIGHT LECTURES

FOR 1894

DELIVERED BEFORE THE ALUMNI ASSOCIATION OF THE COLLEGE
OF PHYSICIANS AND SURGEONS OF NEW YORK.

BY

R. H. CHITTENDEN, Ph.D.

Professor of Physiological Chemistry in Yale University

NEW HAVEN, CONN.:

TUTTLE, MOREHOUSE & TAYLOR, PUBLISHERS.

1895

PREFACE

The present volume, as explained by the title, consists mainly of a reprint of the Cartwright Lectures for 1894. These lectures were originally printed in the current numbers of the Medical Record, but so many requests have been made for their publication in a more convenient and accessible form that they are now re-issued, through the courtesy of the publishers of the Record, in book-form.

It is hoped that these lectures may prove of value not only in calling attention to some of the fundamental chemico-physiological facts of digestion, but in stimulating closer investigation of the many questions which are so intimately associated with a proper understanding of the processes concerned in the digestion and utilization of the proteid food-stuffs.

R. H. CHITTENDEN.

CONTENTS.

LECTURE I.

The general nature of Proteolytic Enzymes and of Proteids.

LECTURE II.

*Proteolysis by pepsin-hydrochloric acid, with a considera-
tion of the general nature of proteoses and peptones.*

LECTURE III.

Proteolysis by trypsin—Absorption of the main products of proteolysis.

DIGESTIVE PROTEOLYSIS

LECTURE I.

THE GENERAL NATURE OF PROTEOLYTIC ENZYMES AND OF PROTEIDS.

INTRODUCTORY.

In digestive proteolysis we have a branch of physiological study which of late years has made much progress. Chemistry has come to the aid of physiology and by the combined efforts of the two our knowledge of the digestive processes of the alimentary tract has been gradually broadened and deepened. That which at one time appeared simple has become complex, but increasing knowledge has brought not only recognition of existing complexity, but has enabled us, in part at least, to unravel it.

By digestive proteolysis is to be understood the transformation of the proteid food-stuffs into more or less soluble and diffusible products through the agency of the digestive juices, or more especially through the activity of the so-called proteolytic ferments or enzymes contained therein; changes which plainly have for their object a readier and more complete utilization of the proteid foods by the system.

In selecting this topic as the subject for this series of Cartwright Lectures I have been influenced especially by the opinion that both for the physiologist and the physician there are few processes going on in the animal body of greater importance than those classed under the head of digestion. Further, few processes are less understood than those concerned in this broad question of digestive proteolysis, especially those which relate specifically to the digestion of the various classes of proteid food-stuffs, and to the

2

absorption and utilization of the several products formed. Moreover, the subject has ever had for me a strong attraction as presenting a field of investigation where chemical work can advantageously aid in the advance of sound physiological knowledge; and certainly every line of advance in our understanding of the normal processes of the body paves the way for a better and clearer comprehension of the pathological or abnormal processes to which the human body is subject.

You will pardon me if I specially emphasize in this connection the fact that advance along the present lines was not rapid until physiologists began to appreciate the importance of investigating the chemico-physiological problems of digestion by accurate chemical methods. Something more than simple test-tube study, or even experimental work on animals, is required in dealing with the changes which complex proteids undergo in gastric and pancreatic digestion. The nature and chemical composition of the proteids undergoing digestion, as well as of the resultant products, are necessary preliminaries to any rightful interpretation of the changes accompanying digestive proteolysis; but physiology has been slow to appreciate the significance of this fact, and, until recently, has done very little to remedy the noticeable lack of accurate knowledge regarding the composition and nature of the proteid and albuminoid substances which play such an important part in the life-history of the human organism, either as food or as vital constituents of the physiologically active and inactive tissues. This is to be greatly deprecated, since our understanding of the nature of proteolysis, of the mode of action of the enzymes or ferments involved, and of the relationships of the products formed, is dependent mainly upon an accurate determination of the exact changes in chemical composition which accompany each step in the proteolytic process.

How otherwise can we hope to attain a proper appreciation of the real points of difference between bodies so closely related as those composing the large group of proteids and albuminoids? Surely, in no other way can we measure the nature or extent of the changes involved in the various phases of proteolysis than by a thorough study of chemical composition and constitution, as well as of chemical reactions and general properties.

In the early history of physiology there was, quite naturally, little or no thought given to the nature of proteolytic changes. The gastric juice, as one of the first digestive fluids to be studied, was recognized as a kind of universal solvent for all varieties of food-stuffs, and this even long before anything was known regarding its composition, but beyond this point knowledge did not extend. Active study of the gastric juice, as you well know, dates from 1783, when the brilliant Italian investigator Spallanzani commenced his work on digestion. The names of Carminati, Werner and Montégre[1] are also associated with various phases of work and speculation in this early history of the subject, especially those which pertained to the possible presence of acid in the stomach juices. In 1824, however, Prout showed conclusively that gastric juice was truly acid, and, moreover, that the acidity was due to the presence of free hydrochloric acid, and not to an organic acid. Still, many observations failed to show the presence of an acid fluid in the stomach, and it was not until Tiedemann and Gmelin's[2] masterly researches were published that the cause of this discrepancy was made clear. It was then seen that the secretion of an acid gastric juice was dependent upon stimulation or irritation of the mucous membrane of the

[1] See Berzelius's Lehrbuch der Chemie, Band 9, p. 205, 4te Auflage, for an account of these early discoveries.
[2] Tiedemann und Gmelin : Die Verdauung nach Versuchen. Heidelberg und Leipzig. 1826.

stomach, and that so long as the stomach was free from food or other matter capable of stimulating the mucosa, it contained very little fluid, and that neutral or very slightly acid in reaction. These early observers also recorded the fact that the amount or strength of acid increased with the outpouring of the secretion, incidental to natural or artificial stimulation, thus giving a hint of the now well-known fact that any and every secretion may show variations in composition incidental to the character and extent of the stimulation which calls it forth.

The period between 1825 and 1833 was characterized especially by the presentation of the many results bearing on gastric digestion obtained by Dr. Beaumont on Alexis St. Martin, followed a little later, in 1842, by a long period of experimentation by many physiologists, as Blondlot,[1] Bassow,[2] Bardeleben,[3] Bernard,[4] Bidder and Schmidt,[5] and many others on methods of establishing gastric fistulæ on animals, by which many interesting results were accumulated regarding the physiology of gastric digestion. Up to 1834, however, there was no adequate explanation offered of the solvent power of the stomach juice; aside from the presence of hydrochloric acid, nothing could be discovered by the earlier chemists to account for the remarkable digestive action. Eberle,[6] however, attributed to the mucous membrane of the stomach a catalytic action, and claimed that it only needed the presence of a small piece of the stomach mucosa with weak hydrochloric acid for the manifestation of solvent or digestive power. It remained for Schwann,[7] to show the true explanation of this phenomenon,

[1] Traité analytique de la Digestion. Paris, 1842.
[2] Bulletin de la Société des Naturalistes de Moscou, vol. 16. 1842.
[3] Archiv für physiol. Heilkunde, vol. 8. 1849.
[4] Lecons de Physiologie de la Digestion. Paris, 1867.
[5] Die Verdauungssäfte.
[6] Physiologie der Verdauung. Würzburg, 1834.
[7] Ueber das Wesen der Verdauungsprocesse. Müller's Archiv, 1836, p. 90.

and although he was unable to make a complete separation
of the active principle which he plainly believed existed,
he gave to it the name of pepsin. Wassmann, Pappenheim,[1]
Valentin, and later Elsässer,[2] all endeavored to obtain the
substance in a pure form, and Wassmann,[3] in 1839, surely
succeeded in obtaining a very active preparation of the ferment—one capable of exerting marked digestive action
when mixed with a little dilute acid.   Thus, a true understanding of the general nature of gastric juice was finally
arrived at, and the cause of its digestive power was right-
fully attributed to the presence of the ferment pepsin and
the dilute acid.   Further, the analysis of human gastric
juice made by Berzelius,[4] in 1834, showed that the secretion
contains very little solid matter (1.26 per cent.), thus calling attention to the fact that the digestive power of this
fluid is out of all proportion to the amount of pepsin, or
even to the amount of total solid matter present, and
consequently paving the way for a general appreciation of
the peculiar nature of the dominant body, *i.e.*, the pepsin.
The original conception regarding the manner in which
gastric juice exerts its solvent power on proteid foods was
apparently limited to simple solution ; chemical solution if
you choose, brought about by catalytic action, but without
any hint as to the possible nature of the soluble products
formed.   Mialhe,[5] however, recognized the fact that this
transformation, by which insoluble and non-diffusible proteid matter was converted into a soluble and diffusible product, was a form of hydration, comparable to the change of
insoluble starch into soluble sugar, and he named the
hypothetical product albuminose.   Mialhe's study of the

[1] Zur Kenntniss d. Verdauung.   Breslau, 1839.
[2] Magenerweichung der Säuglinge.   Stuttgart und Tübingen, 1846.
[3] Lehmann's Lehrbuch d. physiol. Chem., Band 2, p. 41, 2te Auflage.
[4] Lehrbuch der Chemie, Band 9, p. 209.
[5] Canstatt's Jahresbericht d. Pharm., 1846, p. 163.

matter in 1846 was followed by Lehmann's [1] investigation of the subject, and the coining of the word peptones as an appropriate name for the soluble products of gastric diges- tion. The peptones isolated by Lehmann were described as amorphous, tasteless substances, soluble in water in all pro- portions and insoluble in alcohol. They were likewise pre- cipitated by tannic acid, mercuric chloride, and lead acetate, and were considered as weak acid bodies, having the power of combining with bases to form salts of a more or less indefinite character. Twelve years later, in 1858, Mulder [2] gave a more complete description of peptones, but his study of the subject failed to advance materially our knowledge of the broader questions regarding the nature of the proc- ess, or processes, by which the so-called peptones were formed. A year later, in 1859, Meissner [3] brought forward the first of his contributions, and during the following three or four years several communications were made representing the work of himself and pupils upon the ques- tion of gastric digestion, or more especially upon the character of the products resulting from the digestive action of pepsin-hydrochloric acid.

The general tenor of Meissner's results is shown in the description of a row of products as characteristic of the proteolytic action of pepsin-acid on proteid matter. In other words, there was a clear recognition of the fact that proteid digestion in the stomach, through the agency of the ferment pepsin, is something more than a simple conver- sion of the proteid into one or two soluble products. The several bodies then isolated were named parapeptone, metapeptone, dyspeptone, α, β, and γ peptone ; names now seldom used, but significant as showing that at this early

[1] Lehmann's Physiologische Chemie, Band 2, p. 318.
[2] Archiv f. d. Holländ. Beitr., 2, 1858.
[3] Zeitschr. f. rat. Med., Band 7, 8, 10 und 14.

date there was a full appreciation of the fact that digestive proteolysis as accomplished by the ferment pepsin is an intricate process, accompanied by the formation of a series of products which vary more or less with the conditions under which the digestion is conducted.

This was the commencement of our more modern ideas regarding digestive proteolysis, but only the commencement, for it ushered in an era of unparalleled activity, in which Brücke, Schützenberger, and Kühne each contributed a large share toward the successful interpretation of the results obtained. Further, knowledge regarding the proteid-digesting power of the pancreatic juice was rapidly accumulating, thus broadening our ideas regarding digestive proteolysis in general. Corvisart[1] had called attention to the proteolytic power of the pancreatic juice in 1857, and although his observations were more or less generally discredited for a time, they were eventually confirmed by Meissner,[2] Schiff, Danilewsky,[3] and Kühne,[4] the latter particularly contributing greatly to the development of our knowledge concerning this phase of digestive proteolysis. The proteolytic power was proved to be due to a specific ferment or enzyme, now universally called trypsin, which digests proteid foods to the best advantage in the presence of sodium carbonate. Digestive proteolysis in the human body was thus shown to be due mainly to the presence of two distinct enzymes, the one active in an acid fluid, the gastric juice, the other in an alkaline-reacting fluid, the pancreatic juice, but both endowed with the power of digesting all varieties of proteid foods, with the formation of a large number of more or less closely related products.

[1] Sur une Fonction peu connue du Pancréas : La digestion des aliments azotés. Paris, 1857–58.
[2] Verdauung der Eiweisskörper durch den pankreatischen Saft. Zeitsehr. f. rat. Med., 3d ser., Band 7, p. 17. 1859.
[3] Virchow's Archiv, Band 25, p. 267. 1862.
[4] Ibid., Band 39, p. 130. 1867.

So much for the early history of our subject, and now, without attempting any exhaustive sketch of its gradual development during the last decade and a half, allow me to present to you digestive proteolysis as it stands to-day, developed somewhat, I trust, by the results I have been able to contribute to it during the last twelve years.

THE GENERAL NATURE OF PROTEOLYTIC ENZYMES.

These peculiar bodies owe their origin to the constructive power of the gland-cells from which the respective secretions are derived.   During fasting, the epithelial cells of the gastric glands and of the pancreas manufacture from the cell-protoplasm a specific zymogen or ferment-antecedent, which is stored up in the cell in the form of granules. These granules of either pepsinogen or trypsinogen, as the case may be, are during secretion apparently drawn upon for the production of the ferment, and it is an easy matter to verify Langley's[1] observation that the amount of pepsin, for example, obtainable from a definite weight of the gland-bearing mucous membrane is proportionate to the number of granules contained in the gland-cells.   During ordinary secretion, however, these granules of zymogen do not entirely disappear from the cell.   When secretion commences and the granules are drawn upon for the production of ferment, fresh granules are formed, and inasmuch as these latter are produced through the katabolism of the cell-protoplasm it follows that anabolic processes must be simultaneously going on in the cell, by which new cell-protoplasm is constructed.   Hence, as Heidenhain, Langley, and others have pointed out, during digestion there are at least three distinct processes going on side by side in

[1] Proceedings of the Royal Society, vol. 32, p. 20 ; On the Histology and Physiology of the Pepsin-forming Glands.

the gland-cell, viz., the conversion of the zymogen stored up in the cell into the active ferment, or other secretory products, the growth of new cell-protoplasm, and the attendant formation of fresh zymogen to replace, or partially replace, that used up in the production of the ferment. Consequently, we are to understand that in the living mucous membrane of the stomach there is little or no preformed pepsin present. Similarly, the cells of the pancreatic gland are practically free from the ferment trypsin. In both cases the cell-protoplasm stores up zymogen and not the active ferment, but at the moment of secretion the zymogen is transformed into ferment and possibly other organic substances characteristic of the fluid secreted. Absorption of the products of digestion tends to increase the activity of the secreting cells, but we have no tangible proof that any particular kinds of food are directly peptogenous, *i.e.*, that they lead to a storing up in the gastric cells, for example, of pepsinogen, although it may be that the so-called peptogenous foods give rise to a more active conversion of pepsinogen into pepsin.[1] As already stated, the zymogen is manufactured directly from the cell-protoplasm, and the constructive power is certainly not directly controlled by the character of the food ingested.

All this in one sense is to-day ancient history, but I recall it to your minds in order to emphasize the fact that these two energetic ferments or enzymes stand in close relation to the protoplasm of the cell from which they originate. So far as we can measure the transformations involved, there are only two distinct steps in the process, viz., the formation of the inactive zymogen stored up in the cell, and the conversion of the antecedent body into the soluble and active ferment. In this connection Pod-

[1] Langley and Edkins : Pepsinogen and Pepsin. Journal of Physiology, vol. 7, p. 394.

wyssozki[1] has reported that the mucous membrane of the
stomach exposed to the action of oxygen gas shows a
marked increase in the amount of pepsin, from which he
infers that the natural conversion of pepsinogen into pepsin
is an oxidation process. Further, he claims the existence
of at least two forms of pepsinogen in the stomach mucosa,
one closely akin to the ferment itself and very easily
soluble in glycerin, while the other is more insoluble in
this menstruum. Langley and Edkins,[2] however, find
that oxygen has no effect whatever on the pepsinogen
of the frog's mucous membrane, thus throwing doubt on
the above conclusion. Still, Podolinski[3] claims that trypsin originates from its particular zymogen through a
process of oxidation, and Herzen[4] has proved that the
ferment can be reconverted into trypsinogen under the
influence of carbon-monoxide and again transformed into
the ferment by contact with oxygen gas. This latter
observer[5] has also noticed a connection between the
amount of trypsin obtainable from the pancreas and the
dilatation of the spleen, from which he was eventually led
to conclude that the spleen during its dilatation gives birth
to a zymogen-transforming ferment which thus leads to the
production of trypsin, presumably from the already manufactured zymogen. In any event, their peculiar origin
lends favor to the view that these two enzymes are closely
allied to proteid bodies, and that they are directly derived
from the albuminous portion of the cell-protoplasm.
Analysis shows that they always contain nitrogen in fairly

[1] Pflüger's Archiv f. Physiol., Band 39, p. 68.
[2] Journal of Physiol., vol. 7, p. 400.
[3] Pflüger's Archiv f. Physiol., Band 13, p. 422.
[4] Ueber den Rückschlag des Trypsins zu Zymogen unter dem Einfluss
der Kohlenoxydvergiftungen. Pflüger's Archiv f. Physiol., Band 30,
p. 308.
[5] Ueber den Einfluss der Milz auf die Bildung des Trypsins. Pflüger's
Archiv f. Physiol., Band 30, p. 295.

large amount, although the percentage is sometimes less than that found in a typical proteid body.

It must be remembered, however, that in spite of oft-repeated attempts to obtain more definite knowledge regarding the composition of these proteolytic enzymes our efforts have been more or less baffled. We are confronted at the outset with the fact that no criterion of chemical purity exists, either in the way of chemical composition or of chemical reactions. The only standard of purity available is the intensity of proteolytic action, but this is so dependent upon attendant circumstances that it is only partially helpful in forming an estimate of chemical purity. My own experiments in this direction, and they have been quite numerous, have convinced me that it is practically impossible to obtain a preparation of either pepsin or trypsin at all active which does not show at least some proteid reactions. Furthermore, such samples of these two enzymes as I have analyzed have shown a composition closely akin to that of proteid bodies. I will not take time to go into all the details of my work in this direction, contenting myself here with the statement that the purest specimens of pepsin and trypsin I have been able to prepare have always shown their relationship to the proteid bodies by responding to many of the typical proteid reactions, and their composition, though somewhat variable, has in the main substantiated this evident relationship.

The most satisfactory method I have found for obtaining a comparatively pure preparation of pepsin, and one at the same time strongly active, is a modification of the method published some years ago by Kühne and myself.[1] The mucous membrane from the cardiac portion of a pig's stomach is dissected off and washed with water. The upper surface of the mucosa is then scraped with a knife

[1] Zeitschr. f. Biol., Band 22, p. 428.

until at least half of the membrane is removed. These
scrapings, containing the fragments of the peptic glands,
are warmed at 40°C. with an abundance of 0.2 per cent.
hydrochloric acid for ten to twelve days in order to transform all of the convertible albuminous matter into peptone. The solution is then freed from insoluble matter by
filtration and immediately saturated with ammonium sulphate, by which the pepsin, with some albumose, is precipitated in the form of a more or less gummy, or
semi-adherent mass. This is filtered off, washed with a
saturated solution of ammonium sulphate and then dissolved in 0.2 per cent. hydrochloric acid. The resultant
solution is next dialyzed in running water until the ammonium salt is entirely removed, thymol being added to
prevent putrefaction, after which the fluid is mixed with
an equal volume of 0.4 per cent. hydrochloric acid and
again warmed at 40° C. for several days. The ferment is
then once more precipitated by saturation of the fluid with
ammonium sulphate, the precipitate strained off, dissolved
in 0.2 per cent. acid and again dialyzed in running water
until the solution is entirely free from sulphate. The
clear solution of the ferment obtained in this manner can
then be concentrated at 40° C. in shallow dishes, and if
desired the ferment obtained as a scaly residue. So prepared, the pepsin is certainly quite pure, that is comparatively, and although it may contain some albumose, the
latter must be very resistant to the action of the ferment;
indeed, pepsin is in many respects an albumose-like body
itself.

In any event, the enzyme prepared in this manner shows
decided proteid reactions, and contains nitrogen corresponding more or less closely to the recognized composition
of an albumose. My own belief, therefore, is that these
enzymes, both pepsin and trypsin, are proteid bodies

closely related to the albumoses. They are soluble in water and more or less soluble in glycerin ; at least glycerin will dissolve them from moist tissues, or from moist pre-cipitates containing them. Langley,[1] however, states, and perhaps justly, that we have no positive proof that either ferments or zymogens are soluble in pure strong glycerin, and that if they are soluble, it is extremely slowly. In dilute glycerin, however, these ferments dissolve readily, as we very well know. Furthermore, they are practically non-diffusible, and, like many albumoses, are precipitated in part by saturation with sodium chloride and completely on saturation with ammonium sulphate.

When dissolved in water and heated above 80° C., these enzymes are decomposed to such an extent that their pro-teolytic power is totally destroyed. The amount of coagu-lum produced by heat, however, is comparatively small, though variable with different preparations. Thus with trypsin, Kühne originally considered that boiling an aque-ous solution of the ferment would give rise to about twenty per cent. of coagulated proteid and eighty per cent. of peptone-like matter. With the purer preparations now obtainable there is apparently less coagulable matter present, and Loew[2] has succeeded in preparing from the pancreas of the ox a sample of trypsin containing 52.75 per cent. of carbon and 16.55 per cent. of nitrogen, and yield-ing only a small coagulum by heat. Loew considered the ferment to be a true peptone, but in view of our present knowledge regarding the albumoses, I think we are justi-fied in assuming it to be an albumose-like body rather than a true peptone. At the same time it may be well to again emphasize the fact that our only " means of determining

[1] Gamgee's Physiological Chemistry of the Animal Body, vol. 2, p. 4. 1893.
[2] Ueber die chemische Natur der ungeformten Fermente. Pflüger's Archiv f. Physiol., Band 27, p. 203.

the presence of an enzyme is that of ascertaining the change which it is able to bring about in other substances, and since the activity of the enzymes is extraordinarily great, a minute trace suffices to produce a marked effect. From this it follows that the purified enzymes which give distinct proteid reactions might merely consist of very small quantities of a true non-proteid enzyme, adherent to or mixed with a residue of inert proteid material."[1] This quotation gives expression to a possibility which we certainly cannot ignore, but my own experiments lead me to believe firmly in the proteid nature of these two enzymes. Further, we find partial substantiation of this view in the results obtained by Wurtz[2] in his study of the vegetable proteolytic ferment papain, and in my own results from the study of the proteolytic ferment of pineapple juice.[3] Thus, Wurtz prepared from the juice of Carica papaya an active sample of papain, and found it to contain on analysis about 16.7 per cent. of nitrogen and 52.5 per cent. of carbon, while the reactions of the product likewise testified to the proteid nature of the enzyme. Martin, too, has concluded from his study of papain that the ferment is at least associated with an albumose.[4]

With the proteolytic ferment of pineapple juice my observations have led me to the following conclusions, viz., that the ferment is at least associated with a proteid body, more or less completely precipitable from a neutral solution by saturation with ammonium sulphate, sodium chloride, and magnesium sulphate. This body is soluble in water, and consequently is not precipitated by dialysis. It is further non-coagulable by long contact with strong

[1] Sheridan Lea : Chemical Basis of the Animal Body, p. 55.
[2] Sur la papaine : Contribution à l'histoire des ferments solubles, Comptes Rendus, Tome 90, p. 1379. *Ibid.*, Tome 91, p. 787.
[3] On the Proteolytic Action of Bromelin, the Ferment of Pineapple Juice. Journal of Physiol., vol. 15, p. 249.
[4] The Nature of Papain, etc. Journal of Physiol., vol. 6, p. 336.

alcohol, and its aqueous solution is very incompletely precipitated by heat. Placing it in line with the known forms of albuminous bodies it is not far removed from protoalbumose or heteroalbumose, differing, however, from the latter in that it is soluble in water without the addition of sodium chloride. At the same time, it fails to show some of the typical albumose reactions, and verges toward the group of globulins. In any event, it shows many characteristic proteid reactions, and contains considerable nitrogen, viz., 10.46 per cent., with 50.7 per cent. of carbon. Consequently, we may conclude that the chemical reactions and composition of the more typical proteolytic enzymes, both of animal and vegetable origin, all favor the view that they are proteid bodies not far removed from the albuminous matter of the cell-protoplasm.

Further, the very nature of these substances and their mode of action strengthen the idea that they are not only derived from the albumin of the cell-protoplasm, but that they are closely related to it. One cannot fail to be impressed with the resemblance in functional power between the unformed ferments as a class and cell-protoplasm. To what can we ascribe the particular functional power of each individual ferment? Why, for example, does pepsin act on proteid matter only in the presence of acid, and trypsin to advantage only in the presence of alkalies? Why does pepsin act only on proteid matter, and ptyalin only on starch and dextrins? Why does trypsin produce a different set of soluble products in the digestion of albumin than pepsin does? Similarly, why is it that the cell-protoplasm of one class of cells gives rise to one variety of katabolic products, while the protoplasm of another class of cells, as in a different tissue or organ, manifests its activity along totally different lines? The answer to both sets of questions is, I think, to be found in the chemical constitution of the cell-protoplasm on the one hand, and in the constitution of the individual enzymes on the óther. The varied functional power of the ferment is' a heritage from the cell-protoplasm, and, as I have said, is suggestive of a close relationship between the enzymes and the living protoplasm from which they originate. We might, on purely theoretical grounds, consider that these unformed ferments are isomeric bodies all derived from different modifications of albumin and with a common general structure, but with individual differences due to the extent of the hypothetical polymerization which attends their formation.

Whenever, owing to any cause, the activity of the ferment is destroyed, as when it is altered by heat, strong acids, or alkalies, then the death of the ferment is to be attributed to a change in its constitution; the atoms in the molecule are rearranged, and as a result the peculiar ferment power is lost forever. The proteolytic power of these enzymes is therefore bound up in the chemical constitution of the bodies, and anything which tends to alter the latter immediately interferes with their proteolytic action. But how shall we explain the normal action of these peculiar bodies? Intensely active, capable in themselves of producing changes in large quantities of material without being destroyed, their mere presence under suitable conditions being all powerful to produce profound alterations, these enzymes play a peculiar part. Present in mere traces, they are able to transform many thousand times their weight of proteid matter into soluble and diffusible products. All that is essential is their mere presence under suitable conditions, and strangely enough the causative agent itself appears to suffer no marked change from the reactions set up between the other substances.

There are many theories extant to explain this peculiar method of chemical change, but few of them help us to any real understanding of the matter. These enzymes are typical catalytic or contact agents, and by their presence render possible marked changes in the character of the proteid or albuminoid matter with which they happen to be in contact. But the conditions under which the contact takes place exercise an important control over the activity of the ferment. Temperature, reaction, concentration of the fluid, presence or absence of various foreign substances, etc., all play a very important part in regulating and controlling the activity of these two proteolytic enzymes. In fact, as one looks over the large number of data which have gradually accumulated bearing upon this point, one is impressed with the great sensitiveness of these ferments toward even so-called indifferent substances. Their specific activity appears to hinge primarily upon the existence of a certain special environment, alterations of which may be attended with an utter loss of proteolytic power, or, in some less common cases, with a decided increase in the rate of digestive action. This constitutes one of the peculiar features of these proteolytic enzymes; powerful to produce great changes, they are nevertheless subject to the influence of their surroundings in a way which testifies to their utter lack of stability. Furthermore, as you well know, conditions favorable for the action of the one ferment are absolutely unfavorable for the activity of the other, and indeed may even lead to its destruction. Thus, while pepsin requires for its activity the presence of an acid, as 0.2 per cent. HCl, trypsin is completely destroyed in such a medium. Again, trypsin exhibits its peculiar proteolytic power in the presence of sodium carbonate, a salt which has an immediate destructive action upon pepsin. Hence, a medium which is favor-

3

able for the action of the one ferment may be directly antagonistic to the action of the other.

Another factor of great moment in determining the activity of these two enzymes is temperature. That which. is most favorable for their action is 38° to 40° C., and any marked deviation from this temperature is attended by an immediate effect upon the proteolysis. Exposure to a low temperature simply retards proteolytic action, doubtless in the same manner that cold checks or retards other chemical changes. There is no destruction of the ferment, even on exposure to extreme cold, the enzyme being simply inactive for the time being. Exposure of either pepsin or trypsin to a high temperature, say 80° C., is quickly followed by a complete loss of proteolytic power, i. e., the ferment is destroyed. It is to be noticed, however, that the destructive action of heat is greatly modified by the attendant circumstances. Thus, fairly pure trypsin, dissolved in 0.3 per cent. sodium carbonate, is completely destroyed on exposure to a temperature of 50° C. for five to six minutes, while a neutral or slightly acid solution of the pure enzyme is destroyed in five minutes by exposure to a temperature of 45° C. On the other hand, the presence of inorganic salts and the products of digestion, such as albumoses, amphopeptone, and antipeptone, all tend to protect the trypsin somewhat from the destructive effects of high temperatures, so that in their presence the enzyme may be warmed to 60° C. before it shows any diminution in proteolytic power. Alkaline reaction, combined with the presence of salts and proteid, viz., just the conditions existent in the natural pancreatic secretion, constitute the best safeguard against the destructive action of heat, and under such conditions trypsin may be warmed to about 60° C. before it begins to suffer harm. But all this testifies in no uncertain way to the extreme sensitiveness of the ferment to changes in temperature; a sensitiveness which manifests itself not only in diminished or retarded proteolytic action, but terminates in destruction of the ferment when the temperature rises beyond a certain point.

Similarly, pepsin dissolved in 0.2 per cent. hydrochloric acid feels the destructive effect of heat when a temperature of 60° C. is reached. In a neutral solution, on the other hand, destruction of the ferment may be complete at 55° C. Here, too, peptone retards very noticeably the destructive action of heat, especially in an acid solution of pepsin, so that under such circumstances the ferment may not be affected until the temperature reaches 70° C. I have tried many experiments along this line, not only with pepsin and trypsin, but also with many other ferments. We may briefly summarize, however, all that is necessary for us to consider here in the statement that the pure isolated ferments are far more sensitive to the destructive action of heat than when they are present in their natural secretions. This, as stated, is due not only to the reaction of the respective fluids but also to the protective or inhibitory action of the inorganic salts and various proteids naturally present. We may thus say with Biernacki[1] that the purer the ferment the less resistant it is to the effects of heat.

It is thus plain that these enzymes, capable though they are of accomplishing great tasks, are nevertheless exceedingly unstable and prone to lose their proteolytic power under the slightest provocation. When, however, they are surrounded by their natural environment, the acid or alkali of the respective secretion, together with salts and proteids, they then appear more stable; their natural lability becomes for the time being transformed into semi-stability, and the temperature, for example, at which they

[1] Das Verhalten der Verdauungsenzyme bei Temperaturerhöhungen. Zeitschr. f. Biol., Band 28, p. 49.

lose their peculiar power, is raised ten degrees or more. I
have also found the same to be true of the vegetable proteolytic ferments, and also of the amylolytic ferment of
saliva.

The above facts furnish us, I think, a good illustration
of how dependent these proteolytic enzymes are upon the
proper conditions of temperature, to say nothing of other
conditions, for the full exercise of their peculiar power.
Toward acids, alkalies, metallic salts, and many other compounds they are even more sensitive than toward heat, and
much might be said regarding the effects, inhibitory or
otherwise, produced by a large number of common drugs
or medicinal agents on these two ferments. Any lengthy
discussion of this matter, however, would be foreign to our
subject, and I will only call your attention in passing to
one or two points which have a special bearing upon the
general nature of the enzymes. Take, for example, the
influence of such substances as urethan, paraldehyde, and
thallin sulphate on the proteolytic action of pepsin-
hydrochloric acid[1] and we find that small quantities, 0.1 to
0.3 per cent. tend to increase the rate of proteolysis, while
larger amounts, say one per cent., decidedly check proteolysis. Similarly, among inorganic compounds, arsenious
oxide, arsenic oxide, boracic acid, and potassium bromide[2]
in small amounts increase the proteolytic power of pepsin
in hydrochloric acid solution, while larger quantities check
the action of the ferment in proportion to the amounts
added. Again, with the enzyme trypsin, similar results
with such salts as potassium cyanide, sodium tetraborate,
potassium bromide and iodide[3] may be quoted as showing

[1] Chittenden and Stewart : Studies in Physiol. Chem., Yale University.
Vol. 3, p. 64.
[2] Chittenden and Allen, Ibid., vol. 1, p. 76.
[3] Chittenden and Cummins, Ibid., vol. 1, p. 112.

not only the sensitiveness of the ferment toward foreign substances, but likewise its peculiar behavior, viz., stimulation in the presence of small amounts and inhibition in the presence of larger quantities.

Furthermore, we have found that even gases, as carbonic acid and hydrogen sulphide, exert a marked retarding influence on the proteid-digesting power of trypsin. Moreover, while it is generally stated that proteolytic and other enzymes are practically indifferent to the presence of chloroform, thymol, and other like substances that quickly interfere with the processes of the so-called organized ferments, pepsin and trypsin certainly do show a certain degree of sensitiveness to chloroform, and indeed even to a current of air passed through their solutions. Thus, very recently, Bertels[1] and Dubs,[2] working under Salkowski's direction, have called attention to the peculiar behavior of pepsin to chloroform; their results showing, first, that small amounts of this agent tend to increase the proteolytic power of the enzyme, while larger amounts decrease its digestive power. Another interesting point brought out especially by Dub's experiments is the fact that an impure solution of the ferment, viz., an acid extract, for example, of the mucous membrane of the stomach containing more or less albuminous matter, is far less sensitive to chloroform than an acid solution of the purified ferment, thus showing again the protective influence of proteids and other extraneous matters; the latter guarding the enzyme to a certain extent from both the stimulating and inhibitory action of various agents.

Another point to be emphasized just here is that any chemical substance, such as a metallic salt, having a spe-

[1] Ueber den Einfluss des Chloroforms auf die Pepsinverdauung, Virchow's Archiv, Band 130, p. 497.
[2] Der Einfluss des Chloroforms auf die künstliche Pepsinverdauung, Ibid., Band 134, p. 519.

cific action upon proteid matter, will almost invariably interfere more or less with the proteolytic action of these enzymes, both through a direct action upon the soluble ferment itself, and also through an indirect action in modifying or inhibiting the digestibility of the proteid exposed to proteolysis. All of these facts emphasize more or less the proteid-like nature of the enzymes, or at least the carriers of the ferments. It is further very suggestive that the destruction of these enzymes by heat happens to occur at approximately those temperatures which are generally recognized as the coagulation points of ordinary proteids. Moreover, the apparent lack of stability so characteristic of these ferments, their inherent proneness to alteration, their marked susceptibility to every change in environment, all point to large complex molecules, such as we have in proteids and are familiar with in living protoplasm.

Whatever the exact nature of these proteolytic enzymes, they are certainly endowed with the power of transforming relatively large amounts of proteid matter into soluble products, even though they themselves are present in very small quantity. They are derived, as we have seen, from living protoplasmic cells, and we might perhaps, with v. Nägeli[1] and Mayer,[2] consider them as retaining a portion of that molecular motion so characteristic of living protoplasm, by which the equilibrium of the dead food-proteid may be disturbed and thus changes started which result in what we call proteolysis. However this may be, we must look to some phase of catalytic or contact action as the true explanation of this power of proteolysis. At first glance, any explanation or theory involving the use of catalysis seems exceedingly vague and indefinite, and yet

[1] Theorie der Gährung. München, 1879.
[2] Die Lehre von den chem. Fermenten. Heidelberg, 1882.

many illustrations can be given of chemical reactions where the dominating agent evidently acts in this manner. "We call a force catalytic," says the philosopher of Heilbron, "when it holds no communicable proportion to the assumed results of its action. An avalanche is hurled into the valley. . . . A puff of wind or the fluttering of a bird's wing is the catalytic force which has given the signal for, and which is the cause of the wide-spread disaster."[1] In the older theories of catalytic action, the catalytic agent was supposed to remain passive, but not so in the more modern conception of catalysis. The ferment by its presence makes possible certain changes and combinations which could not occur in its absence, at least under the existing circumstances, although all the other conditions might be favorable. The proteids, for example, have a natural tendency to undergo hydration; thus, simple boiling with dilute acid or exposure to the action of superheated water alone,[2] will produce many if not all of the products formed in natural digestive proteolysis. To be sure, they are not formed as readily as in artificial or natural digestion, and there may be some minor points of difference, but still proteolysis can be imitated in this manner. The proteolytic enzymes simply help on this natural tendency of proteid bodies to undergo hydration, and by their presence and action enable it to occur at lower temperatures than it otherwise could, and at the same time render it more rapid and complete. This is not accomplished, however, by simple contact. The enzymes, we may assume, combine in some manner with the proteid undergoing digestion, starting thereby a train of reactions

[1] Quoted from Gamgee's Physiological Chemistry of the Animal Body, vol. 2, p. 7.
[2] Chittenden and Meara. A study of the primary products resulting from the action of superheated water on coagulated egg-albumin. Journal of Physiol., vol. 15, p. 501.

in which the proteid and the water present are the main actors, they being, however, perfectly passive in the absence of the inciting agent, the enzyme. As expressed by Gamgee, "the ferment phenomena resemble those in which there is apparently a periodic synthesis and dissociation of the catalyzing agent, which acts in a similar manner to the agent which explodes a train of gunpowder."

We can find many illustrations among chemical phenomena where one body, even though present in small quantity, acts as a go-between and makes possible an almost indefinite exchange of matter and energy. Take as an illustration, the part played by water in determining the explosion of oxygen and carbon-monoxide gas. Some years ago, Dixon[1] called attention to the fact that a mixture of these two gases when perfectly dry would not explode even by contact with red hot platinum wire. The presence, however, of a small amount of aqueous vapor would at once cause an explosion to occur. In confirmation of this observation, Traube[2] has reported that a flame of carbon-monoxide gas introduced into a perfectly dry atmosphere is at once extinguished. In a moist atmosphere, on the other hand, the flame will continue to burn indefinitely, that is as long as the CO gas is supplied. In these cases, the water, which is so necessary for the appearance of the reaction, and which furnishes a striking illustration of the action of a contact or catalytic substance is not purely passive. To be sure, only a minimal amount is necessary for the combustion of an indefinite amount of carbonic oxide, but the water enters into the reaction itself. It is to be noticed that carbonic oxide and water alone, even at high temperatures, will not react, but in the presence of

[1] Chemical News, vol. 46, p. 151.
[2] Bericht. d. deutsch. chem. Gesellsch., Band 18, p. 1890.

oxygen the water is decomposed with formation of hydrogen-peroxide, thus:

$$CO + 2\ H_2O + O_2 = CO(OH)_2 + H_2O_2.$$

The hydrogen-peroxide thus formed combines with carbonic oxide to form carbonic acid, which in turn is decomposed into the anhydride CO_2, with regeneration of water, the latter being available for further action of the same order:

$$H_2O_2 + CO = CO(OH)_2$$
$$2\ CO(OH)_2 = 2\ CO_2 + 2\ H_2O.$$

Indeed, as can be readily seen from the equations, this may be kept up indefinitely, a small amount of water, $i.\ e.$, the go-between, the catalytic agent, sufficing to accomplish the transformation of almost any amount of carbon-monoxide. This, I think, furnishes an excellent illustration of the way in which catalytic agents, such as the proteolytic enzymes, may be supposed to act. It is truly contact action, but the agent is not purely passive; the enzyme combines with the substance undergoing proteolysis, and the resultant compound thus formed is enabled now to combine with water and undergo hydrolysis, something which could not be accomplished by the proteid and water alone, that is at body temperature. This new and more complex compound is naturally less stable and soon undergoes dissociation or cleavage with a splitting off of the original enzyme for one product, which is thus available for further action of the same order; while, as other products, we find the hydrated and otherwise altered substances coming from the proteid, and whose formation is the ultimate object of the whole process.

The parallelism between this hypothetical action of the proteolytic enzymes and the known reactions in the above combustion of carbonic oxide is certainly very close, and leaves little doubt that this explanation of enzyme action is, in a general way at least, correct. Thus the carbonic oxide, CO, brought in contact with pure, dry oxygen gas (apparently all that is necessary for its direct oxidation into carbonic acid, CO_2), undergoes no change; the burning CO gas is at once extinguished. Evidently, something more is necessary in order to start the process of oxidation. So, too, in proteolysis; the process, as we shall see later on, is essentially one of hydration, but bring the proteid and the water, or acid-water, together and although all the conditions are apparently favorable for hydration there is, as you know, little or no change. But introduce the catalytic agent and immediately the reaction commences. In the case of the burning CO gas in contact with oxygen, the water acting as contact agent makes oxidation possible, enabling the main actors in the transformation to react upon each other. But, as we have seen, the contact agent is something more than a mere looker-on, it becomes for the time being an integral part of the molecule, undergoing change, combining with it and thus making possible the subsequent alterations characteristic of the specific transformation, in which, however, the regeneration of the contact agent is a prominent feature. So, too, with the proteolytic enzymes, pepsin and trypsin, they are the go-betweens, making possible the union of the proteids with water by combining with the proteid molecule and thus paving the way for both hydration and cleavage. In the cleavage of the complex molecule, we have the regeneration of the ferment as a prominent feature, and in proteolysis we understand that the regenerated ferment may act not only upon more of the original proteid, but likewise upon the primary products of its action, thus giving rise eventually to a row of more or less closely related cleavage products. Finally, we can conceive that the enzyme may gradually be affected by the process, that its regeneration may become less complete, and thus digestive power be eventually diminished. Much more might be said in support of the above hypothesis. On the other hand, some objections might be raised against it, but I know of no more reasonable explanation of enzyme action than that here presented, or one which so well accords with all of the known facts concerning the conditions which modify proteolytic action.[1] Thus, the influence of heat, of the products of proteolysis, of acids, alkalies, and various organic and inorganic salts on the action of these digestive enzymes is such as lends favor to the above view rather than opposes it.

THE GENERAL NATURE OF PROTEIDS.

PROTEIDS are confessedly among the most complex bodies the physiologist has to deal with, while at the same time they are perhaps the most important, not only in view of their wide-spread distribution through animal and vegetable tissues, but because of the prominent part they take in the nutrition of the body. The more our knowledge is broadened concerning these varied substances, the more we are impressed with their complexity, and at the same time with the necessity for a more accurate study of both their composition and constitution. Concerning the latter, full fruition of our hopes is probably in the distant future, but every step of advance in this direction adds greatly to our resources in the interpretation of the varied and complex changes characteristic of proteid metabolism.

[1] Compare L. de Jager, Erklärungsversuch über die Wirkungsart der ungeformten Fermente, Virchow's Archiv, Band 121, p. 182. Also Chandelon : Bulletin de l'Academie Royale de Méd. de Belgique, 1887, 1, p. 289.

Every study of proteid decomposition adds something to our store of knowledge, and gives perhaps an added fact available for broadening our deductions.[1] Moreover, the composition and general reactions of the proteids may be investigated with full confidence of obtaining many useful results, which must necessarily be an aid in interpreting the changes accompanying digestive proteolysis.

Take, for example, the single question of peptonization by gastric digestion. What is the nature of the process? Is the proteid transformed into a soluble and diffusible peptone as a result of hydration and cleavage, or is it a transformation which results from a simple depolymerization of the proteid molecule, i.e., are we to consider albumin and peptone as isomeric bodies? These questions, on which physiologists seem loath to agree, can certainly be answered definitely; not, however, by arguments but by careful experimentation, in which the composition of the proteid undergoing digestion must be a necessary preliminary factor, and the composition of the resultant product, or products, a secondary factor of equal importance. Further, the question needs to be answered not with reference to one proteid merely, but with reference to every proteid capable of digestion by either gastric or pancreatic juice. When these questions have been fully answered in this manner, we shall have positive data to deal with, and our conclusions will rest upon a foundation of fact not easily set aside. This is one of the problems upon which I have been at work for some years, and although progress may in one sense be slow, yet it is sure and gives results of no uncertain character.

First, then, let us consider briefly the nature of the proteids whose proteolysis we may be interested in; remem-

---

[1] See Drechsel: Der Abbau der Eiweisstoffe. Du Bois Reymond's Archiv f. Physiol., 1891, p. 248.

bering, however, that in so doing we can merely touch upon the points essential for our purpose. Allow me to say in parenthesis that there is being published in Moscow a work on proteids alone of five volumes, 900 pages each, which it is supposed will constitute an exhaustive treatise of the subject.[1]

If we attempt to classify all of the proteid bodies hitherto discovered and studied we are at once confronted with a problem of no small proportions. So varied are they in their reactions, solubilities, and behavior toward general reagents, so inclined to merge into each other by almost insensible gradations that it becomes an extremely difficult matter to make an arrangement that will satisfy all the requirements of the case. I have to suggest, however, the following classification, which is merely a modification of several existing ones, based primarily upon chemical composition, and solubility in the more common menstruums.

Proteids may first be divided into three main groups as follows:

I. *Simple Proteids.*—Composed of carbon, hydrogen, nitrogen, sulphur, and oxygen, and yielding by decomposition aromatic bodies such as tyrosin, phenol, indol, etc.

II. *Compound Proteids.*—Composed of a simple proteid united to some non-proteid body.

III. *Albuminoids.*—A class of nitrogenous bodies related to and derived from proteids, but differing especially from the latter by great resistance to the ordinary solvents of true proteids.

The individual members of these three groups may be arranged as follows on the basis of solubility, coagulability, etc.:

[1] Die Einheit der Proteinstoffe. Historische u. experimentelle Untersuchungen, by L. Morokhowetz.

I. SIMPLE PROTEIDS.—*A. Soluble in water.—a.* Coagulable by heat, and by long contact with alcohol: Albumins: serum-albumin, egg-albumin, lacto-albumin; myo-albumin, vegetable albumins. *b.* Non-coagulable by heat and by long contact with alcohol. Proteoses:[1] proto-proteoses, deuteroproteoses. Peptones:[1] amphopeptones, antipeptones, hemipeptones.

B. Insoluble in water, but soluble in salt solutions.—a. More or less coagulable by heat. Globulins. 1. Soluble in dilute and saturated NaCl solutions. Vitellins. 2. Soluble in dilute NaCl solutions, but precipitated by saturation with NaCl. Myosins, paraglobulin[2] or serum-globulin, fibrinogen, myo-globulin, paramyosinogen, cell-globulins. *b.* Non-coagulable by heat, soluble in dilute NaCl solution and precipitated by saturation with NaCl. Heteroproteoses.

C. Insoluble in water and salt solutions, soluble in dilute alcohol—Zein, gliadins.

D. Insoluble in water, salt solutions and alcohol; soluble in dilute acids or alkalies.—a. Coagulable by heat when suspended in a neutral fluid. Acid-albumins, alkali-albumins or albuminates. *b.* Non-coagulable by heat when suspended in a neutral fluid. Antialbumids, dysproteoses, glutenins.

E. Insoluble in water, salt solutions, alcohol, dilute acids and alkalies; soluble in strong acids, alkalies, and in pepsin-hydrochloric acid and alkaline solutions of trypsin.—Coagulated proteids, fibrin.[3]

---

[1] Used in the generic sense ; the proteoses including albumoses, globuloses, myosinoses, elastoses, etc., and the peptones the peptone-products formed in the digestion of any or all proteids.

[2] Not completely insoluble in saturated NaCl solution.

[3] In blood-fibrin we have a good illustration of the fact that these divisions are not absolutely exact, since this form of proteid matter, for example, is somewhat soluble in dilute acids and in salt solutions, although requiring a long time for marked solution.

II. Compound Proteids.—*A. Compounds of a proteid (globulin) with an iron-containing pigment, soluble in water and coagulable by heat and alcohol.* Hæmoglobin, oxyhæmoglobin, methæmoglobin, etc.

B. Compounds of proteids with members of the carbohydrate group. ` Insoluble in water ; soluble in very weak alkalies.—a. True mucins. b. Mucoids or mucinoids.

C. Compounds of proteids with nucleic acid. Phosphorized bodies yielding by decomposition metaphosphoric acid. Insoluble in water and in pepsin-hydrochloric acid, but more or less soluble in alkalies.—Nucleins.

D. Compounds of proteids with nucleins. Very soluble in dilute alkalies.—Nucleoalbumins, as casein of milk, and nucleoalbumins of cell-protoplasm and cell-nuclei, etc.

III. Albuminoids.—*A. Soluble in boiling water with formation of gelatin and yielding by decomposition leucin and glycocoll*—Collagen (gelatin).

B. Insoluble in boiling water, and yielding by decomposition much leucin and some tyrosin, together with glycocoll and lysatin. Slowly hydrated by boiling dilute acids and by treatment with pepsin-hydrochloric acid.—Elastin.

C. Insoluble in water, dilute acids and alkalies, also in gastric and pancreatic juice. Yield leucin and tyrosin by decomposition.—Keratin, neurokeratin.

We may now advantageously consider the composition of a few of the more prominent representatives of the individual groups, taking for illustration those bodies which have been most thoroughly studied, and which we may have occasion to refer to in our discussion of proteolysis. I have not included in the table any of the alteration-products of the proteids formed by the action of pepsin-acid, trypsin, or boiling dilute acids, confining myself here simply to those bodies which occur ready-formed in nature.

Substance	C	H	N		S		Ash	Origin	Author
Serum-albumin							{0.57– / 1.84}	Serum from horse bld......	Hammarsten.[1]
Serum-albumin								...al ...dn......	Hammarsten.[1]
Egg-albumin							1.11	Non-coagulat d..	Hammarsten.[1]
Egg-albumin								...w's milk...	Sebelien.[3]
Vegetable-albumin							0.70	Corn or maize...	Chittenden and Osborne.[4]
...din							0.32	Wheat....	Osborne and Voorhees.[5]
Proteose, vegetable							0.79	Hemialbumose, urine...	Kühne and ...dn...
Proteose, vegetable							2.99	...dn or ...dn....	Osborne and Osborne.[4]
Proteose, vegetable							0.25	Wheat...	Osborne and Voorhees.[5]
Proteose, vegetable							1.80	Flax-seed...	Osborne.[7]
Vitellin, spheroidal	.03						2.20	...dn or ...mzt	...dn and ...mat
Vitellin, crystalline	51.63	6.83	17.49	22.26	1.08	1.89	1.20	...dn or ...nizu	...dn and Osborne.[4]
Vitellin, I ... Gms	51.31	6.98	15.84	23.04	1.81	2.28	0.30	Squash-seed....	Chittenden and Hartwell.[9]
Vitellin, crystalline	52.18	6.88	18.80	21.65	1.09	0.84	0	Squash-seed...	...dn and Hartwell.[8]
Vitellin, spherods	51.23	6.83	16.91	22.48	1.10	0.87	0.54	Flax-seed..	Osborne.[7]
Vitellin, crystalline	52.82	7.32	16.17	21.84	0.39	2.42	0.49	Meat...	Osborne and Voorhees.[5]
Vitellin, crystalline	52.68	7.38	17.43	19.60	0.48		0.56	...dn and ... Miel.[8]
Vitellin, semi-crystalline	52.18	6.84	13.62	27.53	1.71		0.03	Castor bean..	Osborne.[10]
Myosin, 13 different samples	52.71	6.80	12.32	31.20	0.84		0.20	Brazil nut...	Osborne.[10]
Myosin, vegetable	52.93	6.42	12.58	31.28	2.42		0.25	...snt ...mt	...dn and Setchell.[8]
Myosin, vegetable, crystalline	55.23	7.60	13.18				1.45	Muscle-tissue..	...dn and ...dn.[8,11]
Paraglobulin	52.72	7.10	15.02	22.78	0.71		0.63	...dn or maize.	Chittenden and Osborne.[4]
Fibrinogen	53.01	7.05	15.65	22.03	0.82		0.10	Oats...	Osborne.[12]
Zein	52.34		15.91	24.41	0.70		0.30	Blood of horse..	Hammarsten.[13]
Gliadin	52.33	7.07					1.75	Blood of horse..	Hammarsten.[14]
Glutenin	51.58	7.21					0.43	...dn or maize.	...dn and ...dn[14]
... proteid	52.68	6.81	16.85	22.78	0.71		0.51	Wheat...	Osborne and Voorhees.[5]
... proteid	53.85	7.27	17.97	24.41	0.30			Oats...	Osborne.[12]
Fibrin	54.71	7.03	16.70	25.13	0.30		0.27	Wheat...	Chittenden and Bolton.[2]
Oxyhaemoglobin	50.30	6.52	16.67	21.79	0.38		0.25	Egg-albumin...	...dn an Medndel.[8]
...			16.81	21.97	4.02		0.56 Fe	Blood of dog...	Hammarsten.[14]
Min	88.4	7.53		23.20	1.87		0.43 Fe	Blood of pig...	Hoppe-Seyler.[15]
Min	47.30			20.46	1.88		0.39 Fe	From small...	Hüfner.[16]
Chondromucoid	49.58	6.97	13.15	22.30			0.33	Submaxilliary gland...	Hammarsten.[17]
...			12.3				0.35	Cartilage...	Hammarsten.[18]
Nuclein	52.96							...dn ...bin	Mörner.[19]
...								Pus....	V. l ...dn[20]
N...	53.30						0.98	...Os milk	Hoppe-Seyler.[21]
Gelatin	48.41						1.26	...w's milk..	I ...dn[22]
Elastin	49.38						0.90	Connective tissue..	Lillenfeld.[24]
Elastin	54.24						0.72	Neck-band...	...dn and Solley.[25]
Keratin	53.95						1.01	Aorta....	...dn and Hart.[26]
N... ...dn	49.45						1.35	White r ...dn's hair	Schwarz.[27]
N... ...dn	56.99						2.27	...dn brain...	Kühne and ...dn[28]
Reticulin	52.88						0.34	Reticular ssue..	Siegfried.[30]

[1] Jahresbericht. ...f.
[2] Studies in Physiol. Chemistry, Yale U ...iver., vol. 2, p. 126.
[3] Zeitschr. ...
[4] ...
[5] Ibid., vol. 15, p. 463.
[6] Zeitschr. f. Biol. Band 10, p. 109.

* Many of these results ...nt the average of a ...ge number of individual analyses.

[12] Fourteenth Annual Report ... Sta., 1890; 2d paper, Amer. ...nal, vol. 14, p. 212.
[13] Pflüger's Archiv f. Physiol., Band 22, p. 489.
[14] Ibid., Band 22, p. 479.
[15] Hoppe-Seyler's Med. Chem. Untersuch., p. 189.
[16] ...

[22] Zeitschr. f. p...l. Chem., ...d 7, p. 269.
[23] Studies in Physiol. Chemistry, Yale Univer sity, ol. 2, p. 172.
[24] Du Bois Reymond's Archiv f. Physiol., ...8, p 170.
[25] Journal of Physiology, ol. 12, p. 23.
[26] Studies in ...

In considering the results tabulated above, it is to be remembered that all of these bodies, with the exception of keratin, neurokeratin, and reticulin, are more or less digestible in either gastric or pancreatic juice, or indeed in both fluids. I will not take time here to point out the obvious genetic relationships and differences in composition shown by the above data, but will immediately call your attention to the fact that there are other and more important points of difference between many of these proteids which are hidden beneath the surface, and which a simple determination of composition will not bring to light. I refer to the chemical constitution of the bodies, to the way in which the individual atoms are arranged in the molecule, on which hinges more or less the general properties of the bodies and which in part determines their behavior toward the digestive enzymes, as well as toward other hydrolytic agents. These differences in inner structure can only be ascertained by a study of the decomposition products of the proteids, and of the way in which the complex molecules break down into simpler. The nature of the fragments resulting from the decomposition of a complex proteid molecule, gives at once something of an insight into the character of the molecule. Thus, egg-albumin exposed to the action of boiling dilute sulphuric acid yields, among other fragments, large quantities of leucin and tyrosin, the latter belonging to the aromatic group and containing the phenyl radical. Collagen, or gelatin, on the other hand, by similar treatment fails to yield any tyrosin or related aromatic body, but gives instead glycocoll or amido-acetic acid, in addition to leucin, lysin, and other products common to albumin. Its constitution, therefore, is evidently quite different from that of albumin, but the composition of the body reveals no sign of it. Further, we have physiological evidence of this difference

in constitution in that gelatin, though containing even
more nitrogen than albumin, is not able to take the place
of the latter in supplying the physiological needs of the •
body; its food-value is of quite a different order from that
of albumin.

But while all of the individual proteids show many
points of difference, either in composition, constitution,
reactions, or otherwise, they are nearly all alike in their
tendency to undergo hydrolytic decomposition under
proper conditions; the extent of the hydrolysis and
accompanying cleavage being dependent simply upon the
vigor or duration of the hydrolytic process.

Furthermore, all of the simple proteids, at least, give
evidence of the presence of two distinct groups or radicals,
which give rise by decomposition or cleavage to two distinct classes of products. These two groups, which we
may assume to be characteristic of every typical proteid,
Kühne has named the anti- and hemi-group respectively.
This conception of the proteid molecule is one of the
foundation-stones on which rest some of our present theories regarding the hydrolytic decomposition of proteids,
especially by the proteolytic enzymes. Moreover, it is not
a mere conception, for it has been tested so many times by
experiment that it has seemingly become a fact. The two
groups, or their representatives, can be separated, in part,
at least, by the action of dilute sulphuric acid (three per
cent.) at 100° C. Thus, after a few hours' treatment of
coagulated egg-albumin, about fifty per cent. of the proteid
passes into solution, while there remains a homogeneous
mass, something like silica in appearance, insoluble in
dilute acid, but readily soluble in dilute solutions of sodium
carbonate. This latter is the representative of the anti-
group, originally named by Schützenberger[1] hemiprotein,

----

[1] Recherches sur l'albumine et les matières albuminoides. Bulletin de
la Société chimique de Paris, vols. 23 and 24.

but now called antialbumid.[1] It is only slightly digestible in gastric juice, but is readily attacked by alkaline solutions of trypsin, being converted thereby into a soluble peptone known as antipeptone. In the sulphuric acid solution, on the other hand, are found the representatives of the hemi-group; viz., albumoses, originally known as one body, hemialbumose,[2] together with more or less hemipeptone, leucin, tyrosin, etc.

The fact that we have so many representatives of the hemi-group in this decomposition is significant of the readiness with which the so-called hemi-group undergoes change. All of its members are prone to suffer hydration and cleavage, passing through successive stages until leucin, tyrosin, and other simple bodies are reached. These, and other similar crystalline bodies, are likewise the typical end-products of proteolysis by trypsin, and presumably come directly from the breaking-down of hemipeptone. Antipeptone, on the other hand, is incapable of further change by the proteolytic ferment trypsin. Hence, the hemi-group can be identified by the behavior of the body containing it toward trypsin ; i. e., it will ultimately yield leucin, tyrosin, and other bodies of simple constitution to be spoken of later on. The anti-group, however, will show its presence by a certain degree of resistance to the action of trypsin, antipeptone being the final product of its transformation by this agent; i. e., leucin, tyrosin, etc., will not result. In this hydrolytic cleavage of proteids the anti-group does not always appear as antialbumid. It may make its appearance in the form of some related body, the exact character of the product being dependent

---

[1] Kühne : Weitere Mittheilungen über Verdauungsenzyme und die Verdauung der Albumine. Verhandl. d. Naturhist. Med. Ver. zu Heidelberg, Band 1, p. 236.
[2] Kühne und Chittenden : Ueber die nächsten Spaltungsproducte der Eiweisskörper. Zeitschr. f. Biol., Band 19, p. 159.

in great part upon the nature of the hydrolytic agent, but in every case the characteristics of the anti-group will come to the surface when the body is subjected to the action of trypsin.

The above-described treatment of a coagulated proteid with water containing sulphuric acid evidently induces profound changes in the proteid molecule. The conditions are certainly such as favor hydration, and in the case of complex molecules, like the proteids, cleavage might naturally be expected to follow. Analysis of antialbumid from various sources plainly shows that its formation is accompanied by marked chemical changes. Thus, the following data, showing the composition of antialbumid formed from egg-albumin and serum-albumin by the action of dilute sulphuric acid at 100° C., gives tangible expression to the extent of this change:

	Egg-albumin.	Antialbumid[1] from egg-albumin.	Serum-albumin.	Antialbumid[1] from serum-albumin.
C	52.33	53.79	53.05	54.51
H	6.98	7.08	6.85	7.27
N	15.84	14.55	16.04	14.31

In both cases there is a noticeable decrease in nitrogen, and a corresponding increase in the content of carbon. Evidently, then, this cleavage of the albumin-molecule into the anti-group on the one hand, and into bodies of the hemi-group on the other, is accompanied by chemical changes of such magnitude that their imprint is plainly visible upon the resultant products; changes which certainly are far removed from those common to polymerization.

[1] Kühne und Chittenden: Zeitschr. f. Biol., Band 19, pp. 167 and 178.

This proneness of proteid matter to undergo hydration and subsequent cleavage is further testified to by the readiness with which even such a resistant body as coagulated egg-albumin breaks down under the simple influence of superheated water at 130° to 150° C. Many observations are recorded bearing on this tendency of proteid matter, but few observers have carried their experiments to a satisfactory conclusion. A recent study of this question in my own laboratory, has given some very interesting results.[1] Thus, coagulated egg-albumin placed in sealed tubes with a little distilled water and exposed to a temperature of 150° C. for three to four hours, rapidly dissolves, leaving, however, an appreciable residue. The solution reacts alkaline, there is a separation of sulphur, and in the fluid is to be found not albumin, but two distinct albumose-like bodies, together with some true peptone, and a small amount of leucin, tyrosin, and presumably other bodies.[2] The albumose-like bodies are in many ways quite peculiar. In some respects they resemble the albumoses formed in ordinary digestion; but in others they show peculiarities which render them quite unique, so that they merit the specific name of atmidalbumoses, as suggested by Neumeister. What, however, I wish to call attention to here is the composition of these albumoses. Prepared from coagulated egg-albumin by the simple action of heat and water, they show a deviation from the composition of the mother-proteid, which plainly implies changes of no slight degree. This is clearly apparent from the following table:

[1] Chittenden and Meara : A Study of the Primary Products Resulting from the Action of Superheated Water on Coagulated Egg-albumin. Journal of Physiology, vol. 15, p. 501.
[2] Compare Neumeister's experiments on blood-fibrin. Ueber die nächste Einwirkung gespannte Wasserdämpfe auf Proteine und über eine Gruppe eigenthümlicher Eiweisskörper und Albumosen. Zeitschr. f. Biol., Band 26, p. 57.

	Coagulated egg-albumin.	Atmidalbumose precipitated by NaCl.	Atmidalbumose precipitated by NaCl + acid.	Deutero-atmidalbumose.	Antialbumid.
C	52.33	55.13	55.04	51.99	53.79
H	6.98	6.93	6.89	6.60	7.08
N	15.84	14.28	14.17	13.25	14.55
S	1.81	1.66	....	0.98	....
O	23.04	22.00	....	27.18	....

Here we see that two of these primary albumoses formed by the action of superheated water, like the previously described antialbumid, show a loss of nitrogen with a marked increase in the content of carbon. Evidently, they are related to the antialbumid formed by the action of dilute acid. They are, however, soluble in water, and in many ways differ from true antialbumid, but there is evidently an inner relationship. The so-called deuteroatmidalbumose shows a still more noticeable falling off in nitrogen and sulphur, while the content of carbon is more closely allied to that of the mother-proteid. The albumose precipitable by sodium chloride, although different from an albumid, evidently comes from the anti-group and is a cleavage product which in turn may undergo further hydration and splitting by continued treatment. The so-called deutero-body, on the other hand, may well be a representative of the hemi-group.[1]

It is not my purpose here to enter into details connected with the action of superheated water on proteids. Such a course would take us too far from our present subject, but I do wish to emphasize the fact that even the most resistant of proteids has an innate tendency to undergo hydration and cleavage, and that even simple heating with water alone, at a temperature slightly above 100° C., is sufficient

[1] Compare Krukenberg, Sitzungsberichte der Jenaischen Gesellschaft für Medicin, etc. 1886.

to induce at least partial solution of the proteid. Further, this solvent action in the case of water and dilute acids, at least, is certainly associated with marked chemical changes. It is not mere solution, it is not simply the formation of one soluble body, but solution of the proteid is accompanied by the appearance of a row of new products, in which the terminal bodies are crystalline substances of simple composition. Further, this conclusion does not rest upon the results obtained from a single proteid, for I have at various times studied also the primary products formed in the cleavage of casein, elastin, zein, and other proteids by the action of hot dilute acid, and in all cases have obtained evidence of the formation of several proteose-like bodies, as well as of true peptones.

By the action of more powerful hydrolytic agents, such as boiling hydrochloric acid to which a little stannous chloride has been added to prevent oxidation, the proteid molecule may be completely broken down into simple decomposition products, of which leucin, tyrosin, aspartic acid, glutamic acid, glucoprotein, lysin, and lysatinin are typical examples.[1] In other words, by this and other methods of treatment, which we cannot take time to consider, we can easily break down the albumin-molecule completely into bodies which, as we shall see later on, are typical end-products of trypsin-proteolysis, and which are far removed from the original proteid. But, as we have seen, even the primary bodies formed in the less profound hydrolysis induced by superheated water, do not show the composition of the mother-proteid. Hydration and cleavage leave their marks upon the products, and thereby we know that solution of the proteid is the result of something more than a mere rearrangement of the atoms in the molecule.

[1] Hlasiwetz und Habermann, Ann. Chem. u. Pharm., Band 169, p. 150. Also Drechsel, Du Bois-Reymond's Archiv f. Physiol., 1891, p. 255.

Further, we are to remember that boiling dilute acid and superheated water tend to produce a cleavage along specific lines; viz., a cleavage into the anti- and hemi-groups of the molecule, and as representatives of these groups we may, in the hydration of every native proteid, look for two distinct rows of closely related substances.

In digestive proteolysis it will be our purpose to show that cleavage of much the same order occurs, not necessarily resulting, however, in the formation of identically the same products, but certainly accompanied with the production of bodies belonging to the hemi- and anti-groups, although they may be less sharply separated from each other than in the cleavage with dilute sulphuric acid.

The body originally described as hemialbumose, and identified as a product of every gastric digestion, is now known to be a mixture of closely related substances ordinarily spoken of as albumoses,[1] or generically as proteoses. These are primary products in the digestion of every form of proteid matter, intermediate between the mother-proteid and the peptone which results from the further action of the proteolytic enzymes. Associated with the hemialbumoses are corresponding antialbumoses, coming from the anti-half of the proteid molecule, and differing from their neighbors, the hemi-bodies, mainly in their behavior toward the ferment trypsin. Thus, we have the counterpart of the many bodies described by Meissner, although now arranged systematically and on the basis of structural and other differences not thought of in his day.

By the initial action of pepsin-acid, proteids are first transformed into acid-albumin or syntonin, then, by the further action of the ferment, this body is changed into the primary proteoses, proto and heteroproteose, of each of which there must be two varieties, a hemi and an anti.

[1] Kühne und Chittenden: Ueber Albumosen, Zeitschr. f. Biol., Band 20, p. 11.

These may then undergo further transformation into what is known as a secondary proteose, viz., deuteroproteose, of which there must likewise be two varieties, corresponding to the hemi- and anti-groups respectively. By continued proteolytic action there results as the final product of gastric digestion peptones ; approximately, an equal mixture of so-called hemipeptone and antipeptone, generally known as amphopeptone. Such a peptone exposed to the proteolytic action of trypsin should obviously break down in part into simple crystalline bodies, leaving a residue of true antipeptone. In truth, this is exactly what does happen when the peptone resulting from gastric digestion is warmed with an alkaline solution of trypsin. The so-called hemipeptone quickly responds to the action of the pancreatic ferment, and is converted into other products, while the so-called antipeptone resists its action completely, thus giving results in harmony with our general conception of the proteid molecule.

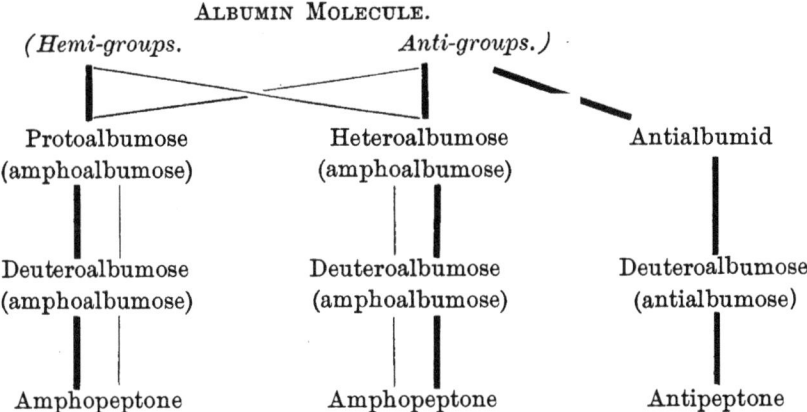

ALBUMIN MOLECULE.

(*Hemi-groups.*   *Anti-groups.*)

| Protoalbumose | Heteroalbumose | Antialbumid |
| (amphoalbumose) | (amphoalbumose) | |

| Deuteroalbumose | Deuteroalbumose | Deuteroalbumose |
| (amphoalbumose) | (amphoalbumose) | (antialbumose) |

| Amphopeptone | Amphopeptone | Antipeptone |

On the basis of these facts, and others not yet mentioned, we may accept provisionally, at least, the above schematic view, suggested in part by Neumeister,[1] of the

[1] Zur Kentniss der Albumosen, Zeitschr. f. Biol., Band 23, p. 391.

general line of proteolysis as it occurs in pepsin-digestion; a view which clearly expresses the significant relationship of the hemi- and anti-groups in the proteid molecule.

The dark and light lines in this scheme are intended to represent the relative share which the hemi- and anti-groups take in the formation of the individual bodies. Thus, we see that protoproteoses have their origin mainly in the hemi-groups of the molecule, although, as the fine line indicates, anti-groups are somewhat concerned in their construction. Heteroproteoses, on the other hand, come mainly from the anti-groups, but still some hemi-groups have a part in their structure. As previously stated, these two primary proteoses by further hydrolytic action may be transformed into secondary products ; viz., into deutero-proteoses, but, as the above scheme indicates, the two deutero bodies will be more or less unlike in their inner nature. In one sense, they are both amphodeuteropro-teoses, but they necessarily differ in the proportion of hemi- and anti-groups they contain. By the still further action of pepsin-acid, the deutero bodies may be changed, in part at least, into peptone, i. e., into amphopeptone, although, as Neumeister has pointed out, protoproteose tends to yield an amphopeptone in which the hemi-groups predominate, while the peptone coming from heteropro-teose contains an excess of anti-groups. Moreover, in the gastric digestion of any simple proteid a certain number of anti-groups are split off in the form of antialbumid, a body which is only slowly digestible in pepsin-acid. By the very powerful proteolytic action of a strong gastric juice, however, antialbumid may be somewhat digested, and is then transformed into antideuteroalbumose, which in turn may be eventually changed into antipeptone.

From these statements it is evident that a given proteid exposed to pepsin-proteolysis may give rise to a large number of products; in fact, to a far larger number than is implied by the names in the above scheme. Thus, at first glance you would be inclined to say there can be only three deuteroalbumoses, for example; one, a pure anti-body, the other two, amphoalbumoses, differing from each other simply in their content of hemi- and anti-groups. It must be remembered, however, that the inner constitution of these bodies, as implied by the relative proportion of the above groups, may vary to almost any extent. Thus, every variation in the number of anti-groups split off from the original albumin molecule to form antialbumid means just so much of a change in the relative proportion of hemi- and anti-groups entering into the structure of both primary and secondary albumoses. Hence, as you can see, digestive proteolysis, even in gastric digestion, is a some-what complex process. We have to deal not only with a number of bodies superficially unlike, as the primary and secondary proteoses and peptones, but these bodies may show marked variations in structure dependent upon the exact conditions attending their formation.

Evidently, the complexities attending digestive prote-olysis are connected primarily with the complex nature of the proteids themselves, while proteolysis, as a process, is made possible through the natural tendency of the proteids to undergo hydration and cleavage.

LECTURE II.

PROTEOLYSIS. BY PEPSIN-ACID.

GASTRIC digestion is essentially an acid digestion. As a
proteolytic agent, pepsin can act only in the presence of
acid, and we have every reason for believing that the
enzyme and the acid form a compound, which in turn combines with the proteid undergoing digestion ; or, what
amounts to much the same thing, that the acid perhaps
forms first a compound with the proteid, to which the pepsin can then unite to form a still more complex compound
capable of undergoing hydration and cleavage. Pepsin-
proteolysis, therefore, is strictly the proteolysis produced
by pepsin-acid. In view of this fact, we may well give a
moment's thought to the nature and origin of this acid.

Without attempting any statement of the gradual development of our knowledge regarding the acid of the gastric
juice, we may accept the now well-established fact that the
acid is hydrochloric acid, and that it has its origin in the
parietal, or so-called border-cells of the gastric glands.
That the acid is derived from the decomposition of chlorides is practically self-evident, but Cahn[1] has added experimental proof which removes all shadow of doubt, through
his study of the gastric secretion in animals deprived for
many days of salt ; the gastric juice in such cases being perfectly neutral in reaction, but normal as regards its content
of pepsin.

[1] Die Magenverdauung im chlorhunger. Zeitschr. f. physiol. Chem.,
Band 10, p. 522.

The way in which the specific gland-cells manufacture free hydrochloric acid out of material contained in an alkaline medium is somewhat doubtful. There are, however, at the present day two theories worthy of special notice. The first is based upon observations made by Maly[1] many years ago, which tend to show that certain mineral salts present in the blood are capable of reacting upon each other with formation of hydrochloric acid. Thus, while the blood is an alkaline fluid, it really owes its alkalinity to the presence of two acid salts, viz., sodium bicarbonate ($HNaCO_3$) and disodium hydrogen phosphate (HNa_2PO_4). This latter compound, acted upon by the carbonic acid of the blood, is transformed into a dihydrogen sodium phosphate with simultaneous formation of acid sodium carbonate, as shown in the following equation :

$$Na_2HPO_4 + CO_2 + H_2O = NaH_2PO_4 + HNaCO_3.$$

This acid sodium phosphate dissolved in a fluid containing sodium chloride, gives rise to free hydrochloric acid by a very simple reaction :

$$NaH_2PO_4 + NaCl = Na_2HPO_4 + HCl.$$

It is also to be noted that the disodium hydrogen phosphate, may, likewise, give rise to hydrochloric acid through its action on calcium chloride, as indicated by the following equation :

$$2 Na_2HPO_4 + 3CaCl_2 = Ca_3(PO_4)_2 + 4NaCl + 2HCl.$$

It is thus evident that hydrochloric acid may originate in the inter-reaction of these several salts which are known to be present in the blood; but obviously, the above reactions cannot take place in the blood itself, and we must look to the selective power of the epithelial cells of the gastric glands, as suggested by Gamgee,[2] for the with-

[1] Untersuchungen über die Quelle der Magensaftsäure. Annalen d. Chem. u. Pharm., Band 173, p. 227.
[2] Physiological Chemistry of the Animal Body, vol. 2, p. 113.

drawal of the needed salts from the blood. Once present in the acid-forming cells, and perhaps aided by the inher-ent qualities of the protoplasm, the necessary chemical reactions may be assumed to take place, after which the newly formed acid may pass from the gland-cells into the secretion of the gland.

A later theory regarding the formation of the acid of the gastric juice emanates from Liebermann.[1] This investi-gator claims the existence in the mucous membrane of the stomach of an acid-reacting, nuclein-like body, which is apparently a combination of the phosphorized substance lecithin with a proteid. To this compound body Lieber-mann gives the name of lecithalbumin. It is apparently located in the nuclei of the gastric cells, is strongly acid in reaction, and, according to Liebermann, is an important agent in the production of the free hydrochloric acid of the gastric juice, although its action is somewhat indirect. According to this theory, the free acid is formed in the mucous membrane of the stomach from sodium chloride, through the dissociating action of the carbonic acid coming from normal oxidation. The thus-formed acid then diffuses in both directions, viz., through the lumen of the gland into the stomach-cavity, and in part in the opposite direc-tion into the veins and lymphatics. It is the assumed function of the lecithalbumin to react with the alkaline sodium carbonate, produced simultaneously with the hydro-chloric acid. This naturally gives rise to the liberation of carbonic acid and to the formation of a non-diffusible sodium-lecithalbumin compound, which is retained for the time being in the body of the cell. When the circulation of the blood, accelerated by the digestive process, returns to its ordinary pace, this latter compound is slowly decom-

[1] Studien über chemische Processe in der Magenschleimhaut. Pflüger's Archiv f. Physiol., Band 50, p. 25. Neue Untersuchungen über das Lecithalbumin. Liebermann, Ibid., Band 54, p. 573.

posed by the carbonic acid with formation of the readily diffusible sodium carbonate, which passes into the blood-current. The rate of this latter reaction is impeded, or, perhaps regulated, by the swelling up of the lecithalbu-min-containing cells, thus rendering the imbibition of the carbonic acid a slow process. The rate of production of the hydrochloric acid by this hypothetical process depends primarily upon the blood supply, and the oxidative changes by which carbonic acid is formed.

There is much that might be said for and against this theory,[1] but we cannot stop to discuss it here. Like the previous theory, it implies the production of hydrochloric acid from a chloride or chlorides, through chemical pro-cesses taking place in the stomach-mucosa, and presumably in the large border-cells of the peptic glands. This hydro-chloric acid, as you know, in the act of secretion, reacts upon the pepsinogen with which it may come in contact, transforming it into pepsin. It also has the power of com-bining with all forms of proteid matter, not excepting the products of proteolytic action, to form acid compounds in which the so-combined acid, although equal quantitatively to the original amount of free acid, is less active in many ways. Thus, it does not possess in the same degree a destructive action on the amylolytic ferments;[2] it does not play the same part in aiding the proteolytic action of pep-sin, and its antiseptic power is far from equal to that of a like amount of free acid.[3]

With relatively large amounts of proteid, we may have half or even quarter saturated proteid molecules, in which

[1] See discussion by Plósz and Liebermann in Jahresbericht für Thier-chemie, Band 22, p. 260.
[2] Chittenden and H. E. Smith : Studies in Physiol. Chem., Yale Uni-ver., vol. i., p. 18.
[3] Compare F. O. Cohn : Ueber die Einwirkung des künstlichen Magensaftes auf Essigsäure- und Milchsäuregährung. Zeitschr. f. physiol. Chem., Band 14, p. 74.

the weakness of the combined acid is far more pronounced than in the case of the fully saturated molecule. Such a condition of things must obviously exist in the early* stages of gastric digestion. With an excess of proteid matter in the stomach, some time must elapse before the secretion of hydrochloric acid will be sufficient to furnish acid for all of the proteid matter present, yet pepsin-prote-olysis does not wait the appearance of free acid. Indeed, the proteid matter may not have combined with more than half its complement of hydrochloric acid before digestive proteolysis is well under way. I have made many analyses of the stomach-contents after test meals, and under other conditions, where no free acid could be detected by the tropaeolin test, or better, by Günzburg's reagent (phlo-roglucin-vanillin), although phenolphthalein as well as lit-mus showed strong acid reaction, and yet not only could acid-albumin be detected in the filtered fluid, but likewise proteoses and peptones. In other words, pepsin-proteolysis can proceed in the absence of free hydrochloric acid, although not at the same pace. Hence, proteoses and even peptones may make their appearance in the stomach-contents at a very early period of digestion, *i. e.*, the final products of proteolysis may be found in a mixture containing even a large proportion of wholly unaltered proteid, and obviously at an early stage in the process. Expressed in other language, a portion of the first formed acid-albumin or syntonin may be carried forward by the digestive process to the secondary proteose and peptone stage, before the larger portion of the ingested proteid food has even combined with sufficient acid to insure the complete formation of acid-albumin. This introduces another factor, to be referred to later on, viz., the relative combining power of different forms of proteid matter, especially the proteoses and peptones, as contrasted with native proteids.

In proof of the statement that pepsin-proteolysis can proceed in the absence of free hydrochloric acid, provided combined acid be present, allow me to cite one or two experiments bearing on this point. A perfectly neutral solution
of egg-albumen, containing 0.8169 gramme of ash-free
albumin per 10 c.c. of fluid, was employed as the proteid
material. In order to completely saturate the proteid contained in 20 c.c. of this neutral albumen solution, 50 c.c. of
0.2 per cent. HCl were required. Two mixtures were
then prepared as follows :

A. Twenty c.c. of the neutral albumen solution + 50 c.c.
0.2 per cent. HCl + 30 c.c. of a weak aqueous solution of
pepsin, perfectly neutral to litmus. This mixture gave
only the faintest tinge of a reaction for free acid when
tested by Günzburg's reagent. *,) : /,.* '

B. Twenty c.c. of the neutral albumen solution + 25 c.c.
0.2 per cent. HCl + 30 c.c. of the neutral pepsin solution.
In this mixture, the proteid matter was obviously only half
saturated with acid.

The two solutions were placed in a bath at 40° C.,
where they were allowed to remain for forty-four hours, a
little thymol being added to guard against any possible
putrefactive changes. At the end of this time the amount
of undigested albumin was accurately determined. The 20
c.c. of original albumen solution contained 1.6338 grammes
of dry coagulable albumin. At the end of the forty-four
hours, *A* contained only 0.5430 gramme of unaltered albumin, or acid-albumin, while *B* contained 1.2225 grammes.
That is to say, in the mixture *A*, where the acid existed
wholly in the form of combined acid, but with the albumin
completely saturated, 1.0908 grammes of the proteid were
converted into soluble albumoses and peptones. In *B*, on
the other hand, where the albumin was only half saturated
with acid, 0.4113 gramme of the proteid was converted

5

into soluble products.   This difference in action is made
more striking by the statement that where the proteid was
only half saturated with acid, 25.1 per cent. of the albumin
was digested; while with a complete saturation of the proteid, 66.7 per cent. of the albumin was digested.

To give emphasis to this matter, a second experiment
may be quoted as follows : The proteid used was the
same neutral solution of egg-albumen containing 0.8169
gramme of albumin per 10 c.c.   Two mixtures were prepared as follows :

A. Ten c.c. of the neutral albumen solution + 21.7 c.c.
0.2 per cent. HCl, the amount needed to completely saturate the proteid, + 40 c.c. of a weak solution of pepsin,
perfectly neutral.

B. Ten c.c. of the albumen solution + 10.9 c.c. 0.2 per
cent. HCl + 40 c.c. of the pepsin solution, making a mixture half saturated with acid.

These two solutions were warmed at 40° C. for seventeen hours.   The extent of digestive action was then determined, when it was found that in A only 0.1638 gramme
of the proteid was undigested, while in B, 0.6088 gramme
remained unaltered.   In other words, where the proteid
was completely saturated with acid, but with an utter lack
of free acid, 79.9 per cent. of the albumin was converted
into albumoses and peptone, while in the mixture half saturated with acid only 25.4 per cent. was digested.

These two experiments thus give striking proof that free
acid is not absolutely essential for pepsin-proteolysis.
Digestion is, to be sure, more rapid and complete when
free hydrochloric acid is present, but proteolysis is still
possible, and even vigorous, when there is a marked deficiency of free acid.   Further, as we have seen, proteolysis
may proceed to a certain extent even though the amount
of acid available is not sufficient to combine with more
than half the proteid matter present.

These facts at once raise the question whether the products of proteolysis may not have a stronger affinity for acid than the native proteids ; an affinity so strong that they may be able to withdraw acid from the acid-albumin first formed. One of our conceptions regarding pepsin-proteolysis is that acid is necessary for every step in the proteolytic process. A primary albumose, for example, cannot be further changed by pepsin, unless there is acid present for it to combine with. This being true, it is clear, in view of the fact that even peptones may appear in a digestive mixture containing an amount of acid insufficient to combine even with the albumin present, that the products of proteolysis must withdraw acid from the acid-albumin first formed. In regard to the first point, my own experiments certainly tend to show that the products of gastric digestion do combine with larger amounts of hydrochloric acid than undigested proteids ; and further, that of the several products of proteolysis, the secondary proteoses combine with a larger percentage of acid than the primary proteoses, while true peptones combine with still larger amounts. In other words, the simpler and more soluble the proteid, the larger the amount of acid it is capable of combining with ; a statement which accords with results obtained by other workers[1] in this direction. Further, another factor of considerable importance in connection with the natural digestive process is that a dissolved proteid, such as protoalbumose for example, will combine more readily with free acid than an insoluble proteid ; from which Gillespie[2] is led to infer that in pepsin-proteolysis where there is no free acid present, only acid-albumin, proteoses may be formed to a limited extent at the expense of some of the acid of the acid-albumin, a portion of the latter being

[1] See especially Gillespie : Gastric Digestion of Proteids. Journal of Anat. and Physiol., vol. 27, p. 207.
[2] Loc. cit.

perhaps reconverted into albumin. The ability of the proteoses, however, to withdraw acid from its combination
with a native proteid is perhaps best indicated by Kossler's[1]
experiments, which show that a solution of acid-albumin
containing only enough hydrochloric acid to hold the albumin dissolved, on being warmed at 40° C. for some hours
with addition of a neutral solution of pepsin, may undergo
partial conversion into albumose or peptone.

In spite of these facts, there is some evidence that while
proteoses and peptones have the power of combining with
more acid than a like weight of native proteid, the latter,
leaving out all action of the pepsin, has a stronger affinity
for the acid; in fact, the firmness or strength of the union
appears to diminish as the products become simpler.[2]
Hence, a peptone separated from a digestive mixture, will
part with its combined acid somewhat more readily than
acid-albumin for example, although on this point there is
not complete unanimity of opinion.[3] In digestive proteolysis, however, where the pepsin is accompanied by a
minimal amount of hydrochloric acid, insufficient perhaps
to even half saturate the proteid present, the formation of
proteoses and peptones must be accompanied by a withdrawal of acid from its combination with the native
proteid.

In illustration of some of these points, and especially of
the statement that the products of gastric digestion have
the power of combining with more hydrochloric acid than
the original proteid, allow me to cite the following experiment: 10 c.c. of a neutral solution of egg-albumen contain-

[1] Beiträge zur Methodik der quantitativen Salzsäurebestimmung im
Mageninhalt. Zeitschr. f. physiol. Chem., Band 17, p. 93.
[2] Compare Blum : Ueber die Salzsäurebindung bei künstlicher Verdauung. Zeitschr. f. klin. Medicin, Band 21, p. 558.
[3] See Sansoni: Beitrag zur kenntniss des Verhaltens der Salzsäure zu
den Eiweisskörpern in Bezug auf die Chemische Untersuchung des
Magensaftes. Berliner klin. Wochenschrift, 1893, Nos. 42 and 43.

ing about 0.82 gramme of pure dry albumin, free from mineral salts, required 23.8 c.c. of 0.2 per cent. hydrochloric acid to completely saturate the proteid matter. A mixture was then prepared as follows: 10 c.c. of the albumen + 24 c.c. 0.2 per cent. HCl + 30 c.c. of a neutral pepsin solution, the mixture showing a faint trace of free acid when tested by Günzburg's reagent. This solution was placed in a thermostat at 38° C., and from time to time a drop of the fluid was removed and tested for free acid. If no reaction could be obtained, 0.2 per cent. hydrochloric acid was added to the mixture, until Günzburg's reagent showed free acid to be again present. The following table shows the rate of disappearance of free acid, and the amounts of 0.2 per cent. HCl required to make good the deficiency. The mixture was placed at 38° C. on February 6th, at 11.30 A.M., and, as stated, contained a trace of free acid, 24 c.c. 0.2 per cent. HCl having been added to accomplish this result.

Time.	Acid added to show trace of free acid.
February 6, 11.30 A. M.	
" 2.15 P. M.	4.5 c.c., 0.2 per cent. HCl.
" 5.00 P. M.	1.0 " " " "
February 7, 8.45 A. M.	3.0 " " " "
" 2.00 P. M.	1.0 " " " "
" 5.00 P. M.	1.5 " " " "
February 8, 8.30 A. M.	1.0 " " " "
" 2.30 P. M.	0.0 " " " "
February 9, 8.30 A. M.	3.0 " " " "
February 10, 9.30 A. M.	2.0 " " " "
	17.0

From these results several interesting conclusions may be drawn, in conformity with the statements already made. Thus, as soon as proteolysis commences, the products formed begin to show their greater affinity for acid by withdrawing acid from its combination with the native proteid, a supposition which is necessary to account for even the starting of the proteolytic process. Further, it is evident that proteoses and peptones combine with a far larger equivalent of acid than the native albumin is capable of; 17 c.c. of 0.2 per cent. HCl being required in the above experiment to satisfy the greater combining power of the newly formed products. This doubtless depends upon the cleavage of the large proteid molecule into a number of smaller or simpler molecules, each of the latter, perhaps, combining with a like number of HCl molecules. This view of the relationship of the individual proteoses and peptones is one more or less generally held, and is supported by many facts.[1] However this may be, it is evident that the products of pepsin-proteolysis combine with a larger amount of hydrochloric acid than the mother-proteid, and that the transformation of the latter, at least under the conditions of this experiment, is a slow and gradual process. In the living stomach, on the other hand, where the secretion of acid is progressing with ever-increasing rapidity, it is easy to see that the process of proteolysis would naturally be much more rapid.

Just here we may recall the theory advanced by Richet[2] quite a number of years ago that the acid of the gastric juice is a conjugate acid, composed of leucin and hydrochloric acid, a theory which has found little acceptance. Klemperer,[3] however, assumed that solutions of leucin hydrochloride with pepsin would not digest albumin, but Salkowski and Kumagawa[4] have shown by experiments that leucin and other amido-acids, as glycocoll, may be dis-

---

[1] See Gillespie : On the Gastric Digestion of Proteids. Journal of Anatomy and Physiology, vol. 27, p. 209.

[2] Le suc gastrique chez l'homme et les animaux, ses propriétés chimiques et biologiques. Paris, 1878.

[3] Zeitschr. f. klin. Medicin, Band 14, Heft 1 and 2.

[4] Ueber den Begriff der freien und gebundenen Salzsäure im Magensaft. Virchow's Archiv, Band 122, p. 235.

solved in hydrochloric acid in such proportion that the solution is practically composed of leucin hydrochloride, without interfering with the digestive action of pepsin-acid on blood-fibrin; the solution being physiologically active, although Günzburg's reagent shows an entirely negative result for free acid. If the matter is studied quantitatively, however, it will be found that the amido-acids combining in this manner with the hydrochloric acid of the gastric juice do give rise to some disturbance of proteolytic action;[1] i. e., digestion may be less rapid, especially on egg-albumin, a conclusion which Salkowski[2] has lately confirmed. Still, under such circumstances, digestion does go on and at a fairly rapid rate; hence, if there is a combination between the acid and these organic bodies, as is indicated by Günzburg's reagent, the acid is still active physiologically, even more so than in the compound formed by the interaction of proteid and acid. In other words, many of these neutral organic bodies that may originate in the stomach through fermentative processes, or otherwise, and which tend to combine with the acid of the gastric juice, do not, as a rule, impede pepsin-proteolysis to the same extent that an excess of proteid matter may. In fact, in artificial digestions long continued, pepsin-acid solutions containing considerable leucin, for example, may accomplish as much in the way of digesting proteid matter as the same amount of pepsin-acid without leucin; but the inhibitory action of the amido-acid is there, and may be shown during the first few hours of the experiment, when less proteoses and peptones are formed than in the control experiment without leucin.

It is foreign to our subject to discuss here methods for

[1] Rosenheim : Centralbl. f. klin. Medicin, 1891, No. 39. F. A. Hofman, ibid., No. 42.
[2] Ueber die Bindung der Salzsäure durch Amidosäuren. Virchow's Archiv, Band 127, p. 501.

the detection of so-called free and combined hydrochloric acid in the stomach-contents, or the special significance of such findings in health and disease. I cannot refrain, however, in connection with what has been said above concerning the proteolytic action of pepsin in the presence of combined acid, from saying a word concerning the usual deductions drawn from the absence of free acid in the stomach-contents. As Langermann[1] has recently expressed it, we have methods for discriminating between free and combined acid; we can, moreover, determine the amount of free acid, but is it not equally important to be able to say something definite concerning the amount of combined acid in the stomach-contents? Even in the absence of free hydrochloric acid there may be a sufficient amount of HCl secreted to answer all the purposes of digestion, and yet at no time may there be any free acid present to be detected by the various color-tests ordinarily made use of. I am aware that in ordinary examinations of the stomach-contents after a test meal the results are essentially comparative, and possibly all that are necessary for clinical purposes. What I wish to emphasize, however, is that in order to pass conclusively upon the sufficiency or insufficiency of the gastric secretion, it is wise to know not only the total acidity of the stomach contents and whether there is free acid or not, but to know more about the amount of combined acid present. Thus, there is a natural tendency to divide the fluids withdrawn from the stomach into three groups, viz., those which contain free acid in moderate amount, those which contain free acid in excess, and those in which free acid is entirely absent; but in the latter group, there may be very marked differences in the amount of acid combined with the proteid and other material present. It appears to me that one of the ques-

[1] Virchow's Archiv, Band 128, p 408.

tions to be answered is whether there is sufficient combined HCl present to meet all the requirements for digestion. If there is, that gastric juice may be just as normal as the one containing free mineral acid, and yet, according to our present tendencies, we should be inclined to call the juice containing no free acid abnormal, although there may be sufficient combined acid present to meet all the requirements for digestion. Hence, in examination of the stomach-contents, it is well to consider the use of those methods which tend to throw light upon the amount of combined acid present, for in my opinion it is only by a determination of the total amount of combined acid that we can arrive at a true estimate of the extent of the HCl deficiency. Obviously, in simple clinical examinations of the stomach-contents after a test meal, where proteid matter is not present in large amount, free acid may reasonably be expected to appear after a definite period ; but in any event, it is well to remember that free hydrochloric acid is not absolutely indispensable for fairly vigorous proteolytic action, and that in the presence of moderate amounts of proteid matter a large quantity of acid is required to even saturate the albuminous material.

Consider for a moment the amount of acid a given weight of proteid will combine with, before a reaction for free acid can be obtained. Thus, Blum[1] has stated that 100 grammes of dry fibrin will require 9.1 litres of 0.1 per cent. hydrochloric acid to completely saturate it. Hence, with a daily consumption of 100 grammes of proteid, there would be needed for gastric digestion 4.5 litres of 0.2 per cent. hydrochloric acid daily, and even this would not suffice to give any free acid, assuming that none of the acid is used over again. The results I have already given for egg-albumin tend to show that 1 gramme of pure

[1] Zeitschr. f. klin. Medicin, Band 21, p. 558.

albumin, free from inorganic salts, when dissolved in a
moderate amount of water will combine with about 30 c.c.
of 0.2 per cent. hydrochloric acid. Consequently, on this
basis, 100 grammes of dry egg-albumin will combine with
3 litres of 0.2 per cent. HCl, and not until this amount of
acid has been added to such a mixture will reaction for
free acid be obtained with Günzburg's reagent. Hence
we can easily see, in view of these figures, that the production of hydrochloric acid by the gastric glands may at
times be very extensive, without the stomach-contents
necessarily containing free acid.

While I am by no means willing to agree with Bunge[1]
that the chief importance of the acid of the gastric juice is
its action as an antiseptic, I am decidedly of the opinion
that the lack of free hydrochloric acid in the stomach-contents is more liable to cause disturbance through the consequent unchecked development of bacteria than through
lack of proteolytic action, assuming, of course, the presence
of a reasonable amount of combined HCl. The hydrochloric acid of the gastric juice unquestionably plays a
very important part in checking the growth and development of many pathogenic bacteria, as well as of less
poisonous organisms, which are taken into the mouth with
the food. On all, or at least on nearly all of these organisms, hydrochloric acid exerts a far greater destructive
action when free than when combined with proteid matter.
As Cohn[2] has plainly shown, both hydrochloric acid and
pepsin-hydrochloric acid quickly hinder acetic- and lactic-
acid fermentation, but when the acid is combined with
peptone, for example, it is no longer able to exercise the
same inhibitory influence. It is also important to note

---

[1] Physiologische und Pathologische Chemie, p. 153.
[2] Ueber die Einwirkung des künstlichen Magensaftes auf Essigsäure-
und Milchsäure gährung. Zeitschr. f. physiol. Chemie, Band 14, p. 75.
See also Hirschfeld : Pflüger's Archiv f. Physiol., Band 47, p. 510.

that the lactic-acid ferment is not so sensitive to hydro-chloric acid as the acetic-acid ferment. Consequently, when lactic-acid fermentation is once developed a compara-tively large amount of HCl is required to arrest it. Hence, as we all know, a diminished secretion of hydro-chloric acid renders possible acid fermentation of the stomach-contents, as well as putrefactive changes which would not occur in the presence of free HCl, and which are very incompletely checked when the acid is over-saturated with proteid matter.

Pepsin-proteolysis, however, may proceed, to some extent, at least, even though a small amount only of com-bined acid is present. The combined acid, however, must be hydrochloric acid, if proteolysis is to be at all marked. To be sure, pepsin will act in the presence of lactic acid, as well as in the presence of other organic acids, and inor-ganic acids, likewise, but such action at the best is consid-erably weaker than the action of pepsin-hydrochloric acid.[1]

The ferment pepsin can exert its *maximum* action only in the presence of free hydrochloric acid. There must be sufficient HCl to combine with all of the proteid matter present, and the products of proteolysis as fast as they are formed, if digestion is to be rapid and attended with the formation of a large proportion of the final products of proteolysis. It is under such conditions that our study of pepsin-proteolysis is usually conducted. Further, it is to be remembered that our knowledge of the products of such proteolytic action depends almost entirely upon data accumulated by artificial digestive experiments. In no other way can we be absolutely certain of the conditions under which the proteolysis is accomplished, for it is a significant fact, perhaps plainly evident from what has

[1] Chittenden and Allen : Influence of Various Inorganic and Alka-loidal Salts on the Proteolytic Action of Pepsin-Hydrochloric Acid. Studies in Physiol. Chem., Yale University, vol. 1, pp. 91, 94.

already been said in the preceding lecture, that the character of the products resulting from ordinary proteolysis is
dependent in great part upon the attendant circumstances.
Thus, with a relatively small amount of acid, and perhaps
also of pepsin, the initial products of proteolysis are especially prominent, while with an abundance of both pepsin
and free acid, coupled with long-continued action, the final
products predominate.   Between these two extremes there
are many possible variations, as was, I think, made clear in
the previous lecture.   At the same time, it is to be noticed
that these differences are mainly differences in the *proportion* of the several products, rather than in the nature of
the resultant bodies.

In a general way, the products of pepsin-proteolysis may
be divided into three main groups, viz., bodies precipitated by neutralization and represented mainly by the
so-called syntonin or acid-albumin; bodies precipitated by
saturation of the neutralized fluid with ammonium sulphate
and represented by proteoses; bodies non-precipitable by
saturation with ammonium sulphate and represented by
amphopeptones.   The relationship of the individual products may be clearly seen from the following scheme,
arranged after the plan suggested by Neumeister.

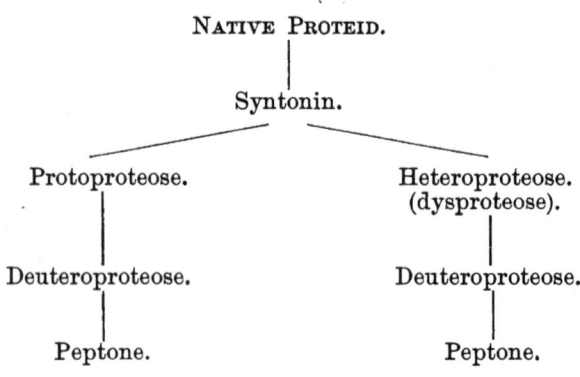

NATIVE PROTEID.

Syntonin.

Protoproteose.   Heteroproteose.
                                        (dysproteose).

Deuteroproteose.   Deuteroproteose.

Peptone.   Peptone.

It is, of course, to be understood that this is not intended to represent anything more than the order of formation of the several bodies, no attention being paid here to the hemi- or anti-character of the several products, or classes of products. Thus, proto and heteroproteose are primary bodies formed directly from the initial product syntonin by the further action of the ferment. In the same sense, deuteroproteose is a secondary proteose, being formed by the further hydration of the primary body. Lastly, peptones, the final products of pepsin-proteolysis, are the result of the hydration and possible cleavage of deuteroproteoses. Further, in almost every gastric digestion there is also formed a small amount of antialbumid, a product insoluble in dilute hydrochloric acid and which consequently appears as an insoluble residue. This body is very resistant to the action of pepsin-acid when once formed, but may be slowly converted, in part at least, into a soluble antialbumose and thence into antipeptone.

All of these bodies can be readily identified in any digestive mixture containing them by a few simple reactions. Thus, after having removed any acid-albumin or syntonin present by neutralization, the concentrated fluid can be tested at once. If primary proteoses are present, the neutral fluid will yield a more or less heavy precipitate on addition of crystals of rock-salt, precipitation being complete only when the fluid is saturated with the salt. Further, if the proteoses are present in not too small quantity, nitric acid added drop by drop to the neutralized fluid will produce a white precipitate, readily soluble on application of heat but reappearing as the solution cools. If primary proteoses are wholly wanting, then no precipitate will be obtained by acid unless the fluid is saturated with salt, in which case a portion of the deuteroproteose will be precipitated. The two primary proteoses differ from each other especially in solubility; protoproteose being readily soluble in water alone, while heteroproteose is soluble only in salt solutions, dilute acids, and alkalies. Hence, when these two bodies are precipitated together by saturation with salt, they may be readily separated by dissolving them in a little dilute salt solution, and dialyzing the fluid in running water until the salt is entirely removed; heteroproteose will then be precipitated, while the proto-body remains in solution.

By long contact with water, and even with concentrated salt solutions, heteroproteose tends to undergo change into a semi-coagulated form, named dysproteose, insoluble in dilute sodium-chloride solutions. This body can be reconverted into heteroproteose, in part at least, by solution in dilute acid, or alkali, and reprecipitation by neutralization.

As a class, the proteoses are characterized by far readier solubility in water than native proteids, by a far greater degree of diffusibility, by non-coagulability by heat and by alcohol, although precipitable by the latter agent. Further, nearly all proteose precipitates are exceedingly sensitive toward heat, tending to dissolve as the fluid is warmed and reappearing as the solution cools. In fact, this peculiarity often serves as a means of identification. Potassium ferrocyanide and acetic acid, picric acid in excess, and likewise cupric sulphate, all precipitate the primary proteoses, while deuteroproteose is only slightly affected by these reagents, or indeed not at all.

In order to separate the secondary proteose from the primary bodies in the absence of peptones, the fluid is neutralized as nearly as possible, and then, after suitable concentration, is saturated with sodium chloride for the partial precipitation of the primary proteoses. To the clear filtrate, acetic acid[1] is added drop by drop as long as a precipitate

---

[1] Saturated with sodium chloride.

results, the latter being composed of a mixture of protoproteose and deuteroproteose. That is to say, protoproteoses
are not completely precipitated from neutral solutions by
saturation with salt alone; a little acid is required to complete it, but this tends to bring down a certain amount of
deuteroproteose. From this filtrate, however, the deutero-
body can be separated in a pure form by dialyzing away
the salt and acid, and then concentrating the fluid and pre-
cipitating with alcohol. When the proteoses are mixed
with peptones, the former must first be separated collect-
ively by saturation of the fluid with ammonium sulphate.

Peptones are especially characterized by non-precipi-
tation with the ordinary precipitants for proteid bodies,
and especially by the fact that they are wholly indifferent
to saturation with ammonium sulphate either in neutral,
acid, or alkaline fluids. This reaction, which constitutes
the main, and perhaps the only absolute method of separating peptones from proteoses must be carried out with great
thoroughness in order to insure a complete precipitation of
deuteroproteose. The latter stands midway between primary proteoses and peptones in many respects, and seems to
share with peptones something of a tendency to resist precipitation by the ammonium salt. Indeed, as Kühne[1] has
recently pointed out, the last traces of deuteroproteose can
be precipitated from the fluid only by long continued boiling of the ammonium sulphate-saturated fluid, and even
then it is seldom complete unless the reaction of the fluid
is alternately made neutral, acid, and akaline, and the heating continued for some time after each change in reaction.
Under such circumstances, the last portions of deuteroproteose separate from the salt-saturated fluid and float on the
surface in the form of an oily or gummy mass, while the

[1] Erfahrungen über Albumosen und Peptone. Zeitschr. f. Biol., Band
29, p. 2.

true peptone remains in the fluid absolutely non-precipitable by the salt.

In this filtrate, peptone can be detected by adding to a small portion of the fluid a very large excess of a strong solution of potassium hydroxide, followed by the addition of a few drops of a very dilute solution of cupric sulphate. If peptone is present a bright red color will appear, the intensity of which, with the proper amount of cupric sulphate, will be proportional to the amount of peptone present. If it is desired to separate the peptone from the ammonium-sulphate-saturated fluid, there are several methods available, of which the following is perhaps the most satisfactory: The fluid is concentrated somewhat, and set aside in a cool place for crystallization of a portion of the ammonium salt. The fluid is then mixed with about one-fifth its volume of alcohol, and allowed to stand for some time, when it separates into two layers—an upper one, rich in alcohol, and a lower one, rich in salts. The latter is again treated with alcohol, by which another separation of the same order is accomplished. Finally, the lighter alcoholic layers containing the peptone are united, and exposed to a low temperature until considerable of the contained salt crystallizes out. The fluid is then concentrated, and after addition of a little water is boiled with barium carbonate until the fluid is entirely free from ammonium sulphate. Any excess of baryta in the filtrate is removed by cautious addition of dilute sulphuric acid, after which the concentrated fluid, reduced almost to a sirupy mass, is poured into absolute alcohol for precipitation of the peptone.

So separated, the peptone formed in gastric digestion is exceedingly gummy, but can be transformed into a yellowish powder, very hygroscopic, of more or less bitter taste, and, when thoroughly dry, dissolving in water with a hissing sound and with considerable development of heat, like phosphoric anhydride.[1]

I have introduced these dry chemical facts, none of which are especially new, because I deem them of considerable importance and because they are not very generally known. In fact, there seems to be a tendency on the part of some who are more or less familiar with the advances made in our knowledge of the products of pepsin-proteolysis to question the existence of these different bodies, or to show at least a spirit of indifference toward these recent facts which have been gradually accumulated, and I may say accumulated at the expense of considerable labor. The time is past for calling the products of gastric digestion peptones ; it is time for a full recognition of the fact that pepsin-proteolysis is synonymous with the production of a row of bodies, chemically and physiologically distinct from each other, each endowed with individuality enough to admit of certain detection, and all bearing a certain specific and harmonious relationship to their neighbors, the other members of the series.

Further, it is not enough to admit the formation of a single intermediate body, midway between syntonin and peptone. The so-called propeptone of the past is simply a mixture of proteoses, of ever changing composition, varying with each change in the proportion of the component proteoses. Each of these proteoses can be detected, under suitable conditions, in the products of every artificial digestion as well as in the stomach-contents, and no better measure of the proteolytic power of the natural stomach-secretion can be devised than a study of the character of the individual bodies present in the stomach-contents after a suitable test meal. The proper tests and separations can

[1] Kühne and Chittenden :   Peptones.   Studies in Physiol. Chem., Yale University, vol. 2, p. 14.

be made with a small amount of the filtered fluid, and much light thrown upon the digestive power of the secretion by even a rough estimate of the proportion of primary and secondary proteoses and peptones formed in a given time, after the ingestion of a certain amount of proteid food.

In pepsin-proteolysis we have to deal, in my opinion, with a series of progressive hydrolytic changes in which peptones are the final products of the transformation. Commencing with the formation of acid-albumin or syntonin, hydrolysis and cleavage proceed hand in hand, under the guiding influence of the proteolytic enzyme, and each onward step in the process is marked by the appearance of a new body corresponding to the extent of the hydrolysis; each body, perhaps, being represented by a row or series of isomers, all externally alike, but different in their inner structure, according to the proportion of hemi- and anti-groups contained in the molecule. As opposed to this theory, we have the older views of Maly,[1] Herth,[2] Henninger[3] and others, based upon observations which tend to show that peptones do not differ in chemical composition from the proteids which yield them. As a matter of fact, the products then analyzed were not peptones at all; they were merely the *primary* products of pepsin-proteolysis, *i. e.*, what we now term primary proteoses, and it is time we stopped using such data to enforce the theory that peptones are polymers of the proteids from which they are derived.

In 1886, the writer, in conjunction with Professor Kühne, commenced a study of the various cleavage products[4] formed by the action of pepsin-hydrochloric acid

---

[1] Ueber die chemische Zusammensetzung und physiologische Bedeutung der Peptone. Pflüger's Archiv f. Physiol., Band 9, p. 585.
[2] Ueber die chemische Natur des Peptones und sein Verhältniss zum Eiweiss. Zeitschr. f. physiol. chem., Band 1, p. 277.
[3] De la Nature et du rôle physiologique des peptones. Paris, 1878.
[4] Globulin and Globuloses. Studies in Physiol. Chem. Yale University, vol. ii., p. 1.

from the better characterized and purer proteids, this being
a continuation of our earlier work on the proteoses and
peptones. formed from blood-fibrin, serum-albumin, etc.
This work I have continued in my laboratory up to the
present time, with many co-workers, and as a result we
have to-day a series of observations gradually accumulated
during these last seven years, some the results of work carried on this last year, which speak in no uncertain way of
the character of both the primary and secondary products
of pepsin-proteolysis. Furthermore, in attempting to settle
this question once for all, I have selected for study examples from the various classes of both animal and vegetable
proteids; and as representatives of the latter have had carried out two lengthy series of experiments on the crystallized proteids which occur so abundantly in some seeds, on
the assumption that these crystalline bodies would furnish
a certain guarantee of purity which might naturally be
lacking in the amorphous proteids of animal origin. Some
of these results are now placed together in the following
tables, a study of which reveals some very interesting facts :

COMPOSITION OF PROTEOLYTIC PRODUCTS FORMED BY
PEPSIN-HYDROCHLORIC ACID.

Proteolysis of Blood-fibrin.

	Mother Proteid.	Protofibrinose.[1]	Heterofibrin-ose.[1]	Deuterofibrin-ose.[1]	Amphopep-tone.[2]
C	52.68	51.50	50.74	50.47	48.75
H	6.83	6.80	6.72	6.81	7.21
N	16.91	17.13	17.14	17.20	16.26
S	1.10	0.94	1.16	0.87	0.77
O	22.48	23.63	24.24	24.65	27.01

[1] Kühne and Chittenden : Zeitschr. f. Biol., Band 20, p. 40.
[2] Kühne and Chittenden : Studies in Physiol. Chem., Yale Univer.,
vol. ii., p. 40.

DIGESTIVE PROTEOLYSIS.

Proteolysis of Paraglobulin.[1]

	Mother Proteid.	Protoglobulose.	Heteroglobulose.	Deuteroglobulose.
C	52.71	51.57	52.10	51.52
H	7.01	6.98	6.98	6.95
N	15.85	16.09	16.08	15.94
S	1.11 }	25.36	24.84	25.59
O	23.24 }			

Proteolysis of Coagulated Egg-albumin.

	Mother Proteid.	Protoalbu-mose.[2]	Heteroalbu-mose.[2]	Deuteroalbu-mose.[2]	Hemipeptone.[3]
C	52.33	51.44	52.06	51.19	49.38
H	6.98	7.10	6.95	6.94	6.81
N	15.84	16.18	15.55	15.77	15.07
S	1.81	2.00	1.63	2.02	1.10
O	23.04	23.28	23.81	24.08	27.64

Proteolysis of Casein from Milk.

	Mother Proteid.	Protocaseose.[4]	Heterocas-cose.[5]	a Deuterocas-eose.[4]	β Deuterocas-eose.[4]
C	53.30	54.58	53.88	52.10	47.72
H	7.07	7.10	7.27	6.93	6.73
N	15.91	15.80	15.67	15.51	15.97
S	0.82 }	22.52	23.18	25.46	29.58
O	22.03 }				

Proteolysis of Myosin from Muscle.[6]

	Mother Proteid.	Protomyosinose.	Deuteromyosinose.
C	52.82	52.43	50.97
H	7.11	7.17	7.42
N	16.77	16.92	17.00
S	1.27	1.32	1.22
O	21.90	22.16	23.39

[1] Kühne and Chittenden : Studies in Physiol. Chem., Yale Univer., vol. ii, p. 12.
[2] Chittenden and Bolton : I*bid.*, vol. ii, p. 153.
[3] Kühne and Chittenden : Zeitschr. f. Biol., Band 19, p. 201.
[4] Chittenden : Studies in Physiol. Chem., Yale Univer., vol. iii., p. 80.
[5] Chittenden and Painter : *Ibid.*, vol. ii., p. 195.
[6] Kühne and Chittenden : I*bid.*, vol. iii, p. 147.

Proteolysis of Elastin.[1]

	Mother Proteid.	Protoelastose.	Deuteroelastose.
C	54.24	54.52	53.11
H	7.27	7.01	7.08
N	16.70	16.96	16.85
S } O }	21.79	21.51	22.96

Proteolysis of Gelatin.[2]

	Mother Proteid.	Protogelatose.	Deuterogelatose.
C	49.38	49.98	49.23
H	6.81	6.78	6.84
N	17.97	17.86	17.40
S	0.71	0.52	0.51
O	25.13	24.86	26.02

Proteolysis of Phytovitellin[3] *(Crystallized) from Squash Seed.*

	Mother Proteid.	Protovitellose.	Deuterovitellose.
C	51.60	51.52	49.27
H	6.97	6.98	6.70
N	18.80	18.67	18.78
S O	1.01 } 21.62 }	22.83	25.25

Proteolysis of Phytovitellin[4] *(Crystallized) from Hemp Seed.*

	Mother Proteid.	Protovitellose.	Deuterovitellose.	Peptone.
C	51.63	51.55	49.78	49.40
H	6.90	6.73	6.73	6.77
N	18.78	18.90	17.97	18.40
S	0.90	1.09	1.08	0.49.
O	21.79	21.73	24.44	24.94

[1] Chittenden and Hart: Studies in Physiol. Chem., Yale Univer., vol. iii, p. 37.
[2] Chittenden and Solley: Journal of Physiol., vol. xii, p. 33.
[3] Chittenden and Hartwell: Ibid., vol. xi, p. 441.
[4] Chittenden and Mendel: Ibid., vol. xvii, p. 48.

Proteolysis of Glutenin[1] from Wheat.

	Mother Proteid.	Protoglutenose.	Heteroglutenose.	Deuteroglutenose.
C	52.34	51.42	51.82	49.85
H	6.83	6.70	6.79	6.69
N	17.49	17.56	17.43	17.57
S	1.08	1.34	1.59	0.80
O	22.26	22.98	22.37	25.09

Proteolysis of Zein.[2]

	Mother Proteid.	Protozeose.	Deuterozeose.
C	55.23	53.29	51.31
H	7.26	6.87	6.88
N	16.13	16.10	16.27
S	0.60	1.54	1.08
O	20.78	22.20	24.46

In considering these results, it is to be noticed that there is a general unanimity of agreement except in the case of the albuminoid gelatin. In the proteolysis of this body, for some reason not explainable, the digestive products show no marked deviation from the composition of the mother-proteid, but in every other instance there is to be traced a distinct tendency toward diminution in the content of carbon, proportional to the extent of proteolysis. In the primary bodies, proto and heteroproteoses, the percentage of carbon is only slightly lowered; indeed, in some few cases, notably in elastin and casein, the primary products show a slight increase in their content of carbon, but in most instances there is a slight falling off in the percentage of this element. In the deuteroproteoses, however, the loss of carbon is very marked. The percentage loss, to be sure, varies with the different proteids, doubtless

[1] Formerly called gluten-casein, and the products gluten-caseoses. Chittenden and E. E. Smith: Journal of Physiol., vol. xi, p. 420.
[2] Chittenden and Williams: Not heretofore published.

dependent in part upon the nature of the proteid itself, and also, I think, upon the strength of the proteolytic agent employed and the duration of the proteolysis. It is to be further noticed that peptones, whenever analyzed, show a still further loss of carbon and also a marked loss of sulphur. In nitrogen there is no constant difference. On the assumption that these various products of proteolysis are formed by a series of hydrolytic changes, accompanied by cleavage of the molecule, we might at first glance look for a marked increase in the content of hydrogen. But when we consider the size of the proteid molecule, with the small proportion of hydrogen contained therein and the large amount of carbon, it is plain that hydrolytic cleavage might naturally leave its mark on the percentage of carbon, rather than on the percentage of hydrogen of the resultant products. In view of these facts, the above results show nothing inconsistent with the theory that pepsin-proteolysis, as a rule, is accompanied by a series of progressive hydrolytic cleavages in which the primary proteoses are the result of a slight hydration, these bodies by continued proteolysis being further hydrated with formation of secondary proteoses, which in turn undergo final hydration and cleavage into true peptones. In accord with this theory, true peptones always show a marked difference in composition from that of the mother-proteid, the most striking feature being the greatly diminished content of carbon, which may be taken as a measure, in part at least, of the extent of the hydrolytic change. And it is to be noticed that the crystallized phytovitellins are no exception to the general rule ; the secondary vitelloses and peptones resulting from proteolysis bear essentially the same relationship to the mother-proteids that the albumoses from egg-albumin do. Moreover, the alcohol-soluble proteids, of which the zein of cornmeal is a good example, show the same general tendency, and it is an interesting fact that the proteoses, or more specifically the zeoses, formed from this peculiar proteid, are readily soluble in water and show the general proteose reactions. It may also be mentioned that these zeoses, as well as the elastoses, are very resistant to further hydrolysis by pepsin-acid, and yield only comparatively small amounts of true peptones.

In connection with this question of the composition of proteoses and peptones as formed by pepsin-proteolysis, it is interesting to note a recent observation recorded by Schützenberger.[1] This experimenter took 350 grammes of moist blood-fibrin, corresponding to 75.5 grammes of dry substance, and subjected it to proteolysis with 2.5 litres of a very strong pepsin-hydrochloric acid solution for five days. The resultant fluid was then freed from acid by treatment with silver oxide, after which the solution was evaporated to dryness on a water-bath and the residue dried *in vacuo*. This residue, termed by Schützenberger fibrin-peptone, was found on analysis to contain 49.18 per cent. of carbon, 7.09 per cent. of hydrogen, and 16.33 per cent. of nitrogen, thus agreeing very closely with true fibrin-peptone as analyzed by Kühne and myself. Further, Schützenberger showed that the fibrin in undergoing this transformation had taken on 3.97 per cent. of water. But to my mind, the most significant fact connected with this experiment is the positive evidence it affords, not only of hydration as a feature of peptonization by pepsin-acid, but that this greatly diminished content of carbon, so characteristic of peptones, and to a less extent of deuteroproteoses, is wholly independent of the methods of separation and purification ordinarily made use of. Thus, Schützenberger, in the above experiment, did not attempt any separation of

---

[1] Recherches sur la constitution chimique des peptones. Comptes Rendus, vol. 115, p. 208.

individual bodies. Proteolysis was carried out under conditions favoring maximum conversion into peptone, and
the resultant product, or products, was analyzed directly
without recourse to any methods of precipitation or purification. To be sure, the substance analyzed could not
have been peptone entirely free from proteose, but in any
event it represented the terminal products of pepsin-proteolysis, and like true amphopeptone contained 3.5 per
cent. less carbon than the original fibrin. Hence, we may
conclude, without further argument, that peptonization in
gastric digestion is the result of distinct hydrolytic action,
in which the original proteid molecule is gradually broken
down, or split apart, into a number of simpler molecules,
the proteoses and peptones.

Peptones, *i. e.*, amphopeptones, are the final products of
gastric digestion; but to how great an extent is actual
peptonization carried on in pepsin-proteolysis? As we
have seen, syntonin, primary proteoses, secondary proteoses, and peptones are all products of pepsin-digestion, and
it might perhaps be assumed that ultimately all of a given
proteid undergoing pepsin-proteolysis would be converted
into amphopeptone. Examination, however, shows that
such is not the case, at least in artificial digestive experiments. Peptones are truly formed, and many times in
large amount, but never under any circumstances have I
been able to effect a complete transformation of any proteid into true peptone by pepsin-proteolysis; there is
always found a certain amount of proteoses more or less
resistant to the further action of the ferment. Obviously,
the nature and proportion of the individual products
formed in any digestive experiment are dependent greatly
upon the attendant conditions; but even with a large
amount of active ferment, an abundance of free hydrochloric acid, a proper temperature, and a long-continued

period of digestion, even five and six days, there is never found a complete conversion into peptone. Indeed, the largest yield of peptone I have ever obtained in an artificial digestion is sixty per cent., while the average of a large number of results under most favorable circumstances is somewhat less than fifty per cent.[1]

We understand that peptones are the products of the hydration and cleavage of previously formed proteoses. The primary proteoses pass into secondary proteoses and these into peptones, but for some reason this transformation after a time becomes a slow and gradual process. At first there is a marked and rapid progression; the proteid undergoing proteolysis is rapidly dissolved, and both proteoses and peptones may be detected in abundance. But if we continue to watch the changing relations of primary and secondary proteoses and peptones, we find that progression soon ceases to be rapid, and eventually travels onward at a snail's pace. Thus, in one experiment with coagulated egg-albumin, there was found at the end of forty-eight hours' digestion with pepsin-hydrochloric acid, only thirty-seven per cent. of peptones with fifty-eight per cent. of proteoses, and yet digestion had been sufficiently vigorous to allow of a complete solution of the proteid in two hours. At the end of seventy-two hours the amount of peptones had increased to about forty-two per cent., the proteoses having correspondingly diminished; but even at the end of seventeen days only fifty-four per cent. of peptones were to be found, thus affording striking evidence of the slow conversion of the first-formed products into peptones.

Naturally, the individual proteoses show marked differences in their rate of conversion into secondary or final

[1] Chittenden and Hartwell : The Relative Formation of Proteoses and Peptones in Gastric Digestion. Journal of Physiol.; vol. xii, p. 12.

products. Take as an illustration some results [1] obtained with caseoses formed in the digestion of the casein of milk. Thus, heterocaseose, a primary product, yielded only fifteen per cent. of peptone after ninety-four hours at 40°C. with a strong pepsin-acid solution. Protocaseose, however, containing some deuterocaseose, under like conditions, yielded thirty-two per cent. of peptone in one hundred and nineteen hours, while pure deuterocaseose gave sixty-six per cent. of peptone in one hundred and thirty-seven hours. Evidently, then, the first-formed soluble products of gastric digestion, *i. e.*, the primary proteoses, are only slowly converted into peptone, since they must first pass through the intermediate stage of deuteroproteose, which is plainly not a rapid process. The deutero-body, on the other hand, once formed is more rapidly converted into peptone, but even this is in no sense a rapid process. Hence, in the artificial digestion of proteids with pepsin-hydrochloric acid, solubility of the proteids may be quite rapid, and even complete in a very short time, but the resultant products will be mainly proteoses and not peptones. The latter are truly formed and in considerable amount, but proteoses, either as primary or secondary bodies, are invariably present and usually in excess of the peptones.

In this connection the question naturally arises how far we are to trust these results in their bearing on the natural process of digestion as it occurs in the living stomach. Obviously, the conditions are quite different in the two cases. In artificial digestions, we have especially the influence of an ever-increasing percentage of soluble products on the activity of the ferment, a condition of things generally considered as more or less inhibitory to enzyme action. We have attempted to measure the real value of

[1] Chittenden and Hartwell, loc. cit., p. 22.

this influence by experiments[1] conducted in parchment dialyzing tubes, in which the conditions are made favorable for the removal of at least some of the products of digestion as fast as they are formed. In these experiments, the dialyzer tubes containing the proteid and pepsin-acid were immersed in a large volume of 0.2 per cent. hydrochloric acid (about three litres), which was gradually changed from time to time, the whole mixture being kept at 40° C. during the entire period of the experiment. The extent of peptonization was then ascertained by analysis of both the contents of the dialyzer tubes and of the surrounding acid, the results being compared with those obtained from control experiments carried on in flasks. Without considering the results in detail, it may be mentioned that the slow and incomplete peptonization so characteristic of artificial gastric digestion is not materially modified by this closer approach to the natural process. The several digestions carried on in the dialyzer tubes were certainly accompanied by a fairly rapid withdrawal of the diffusible products of digestion, yet no noticeable increase in the amount of peptone formed was observed. The results certainly favor the view that the conversion of the primary products of gastric digestion into true peptone is a slow and gradual process, even under the most favorable circumstances, and that this lack of complete peptonization is not due to accumulation of the products of digestion, but is rather an inherent quality of pepsin-proteolysis under all circumstances.

In these dialyzer experiments it was observed that not only did peptones diffuse, but also the proteoses. In fact, it was found that six to eight per cent. of the proteoses

[1] Chittenden and Amerman : A Comparison of Artificial and Natural Gastric Digestion, together with a Study of the Diffusibility of Proteoses and Peptones. Journal of Physiol., vol. xiv, p. 483.

formed passed through the parchment walls of the dialyzer tubes into the surrounding acid in the nine hours' digestion. This led to a study of the diffusibility of proteoses in general, from which we were led to conclude that these bodies possess this power to a greater degree than had hitherto been supposed. As might be expected, it was also found that the attendant conditions modify materially the rate of diffusibility; the two factors especially prominent being temperature and the volume of the surrounding fluid. Thus, 1.9 grammes of protoalbumose dissolved in 200 c.c. of water and suspended in 4.5 litres of water heated to 38° C., diffused through the parchment tube to the extent of 5.09 per cent., while at 10° C. diffusion amounted to only 2.57 per cent. Under somewhat similar conditions, pure peptone diffused to the extent of eleven per cent. in six hours at 38° C. Somewhat singular, however, was the result obtained with deuteroalbumose; this proteose showing a diffusibility considerably less than that of the proto-body. But as Kühne[1] has independently obtained essentially the same results, this apparent anomaly cannot depend upon any errors of work.

It is of course to be understood that diffusion experiments made with dead parchment membranes cannot necessarily be expected to throw much light upon the rate of absorption of these bodies through the living membranes of the stomach and intestine, where, as Waymouth Reid[2] has well said, we have to deal with an absorptive force dependent, no doubt, upon protoplasmic activity, and comparable, in part at least, to the excretive force of a gland-cell. Furthermore, in considering absorption as it occurs in the living stomach, we must necessarily give due weight

[1] Erfahrungen über Albumosen und Peptone. Zeitschr. f. Biol., Band 29, p. 20.
[2] Osmosis Experiments with Living and Dead Membranes. Journal of Physiol., vol. xi, p. 312.

to the selective power of the epithelial cells, a power which may be far more potent even than we suppose. Hence, without attempting at this point to draw any broad deductions from our experiments we may simply lay stress upon the facts themselves, viz., that the primary products of pepsin-proteolysis are diffusible, and, like true peptones, are capable of passing through animal and vegetable membranes, although to a less extent. We may further emphasize the fact that experiments of this character on diffusibility can, at the most, only indicate general tendencies, since every variation in the attendant conditions will exercise some influence upon the final result.

With reference to the bearing digestive experiments made in dialyzer tubes have upon the natural process as carried on in the living stomach, we must necessarily grant that the conditions approximate only in the crudest way to those existent in the alimentary tract. At the same time, if complete peptonization is characteristic of pepsin-proteolysis in the stomach, and failure to obtain such results in an artificial digestion is due to lack of withdrawal of the diffusible products formed, then certainly the experiments carried on in dialyzer tubes, with abundant opportunity for diffusion, and with a large excess of free hydrochloric acid, should show some indications of increased peptone-formation. But none were obtained.

It is more than probable that the rate of absorption of diffusible products from the stomach has been overestimated. Lea,[1] for example, assumes that, " normally the products of digestion, whether proteid or carbohydrate, are never met with in either the stomach or intestine in other than the smallest amounts, frequently to be described as merely traces." This certainly implies a far more rapid absorption of proteoses and peptones from the

[1] Journal of Physiol., vol. xi, p. 240.

stomach than results seem to justify. Indeed, recent facts obtained by Brandl,[1] working under Tappeiner's direction, tend to show that absorption from the stomach is, under some circumstances at least, comparatively slow. Brandl's experiments were conducted on large and vigorous dogs with gastric fistulæ, the stomach being shut off from the intestine by the simple introduction of a small rubber balloon into the pylorus, which when dilated completely closed the orifice. By carefully conducted experiments, it was shown that pure peptone, entirely free from proteose, is absorbed from the empty stomach in proportion to the concentration of the peptone solution. Thus, 7.5 grammes of peptone dissolved in water in such proportion as to make a five per cent. solution, and allowed to remain in the stomach for two hours, lost by absorption only 0.28 gramme, equal to 2.68 per cent. of the peptone introduced. Under similar conditions, a ten per cent. aqueous solution of peptone lost only 4.5 per cent. by absorption. On the other hand, when peptone was introduced in larger quantity, viz., in a twenty per cent. solution, absorption amounted to thirteen per cent. in two hours.

It is thus evident that pure peptones, even when taken into the stomach in fairly large amounts, and under conditions very favorable for rapid absorption, pass into the circulating blood very slowly. Obviously, however, one must not lose sight of the fact that when digestion is under way and the volume of blood consequently increased, there may be a corresponding rise in the rate of absorption. There is perhaps a hint of this conclusion in the influence of alcohol on the absorption of peptone as brought out by some of Brandl's experiments. Thus, it was found that when alcohol was added in considerable quantity to a ten

[1] Ueber Resorption und Secretion im Magen, und deren Beeinflussung durch Arzneimittel. Zeitschr. f. Biol., Band 29, p. 277.

per cent. solution of peptone, the stomach-mucosa was greatly reddened, while in two hours the absorption of peptone amounted to 11.8 per cent. But in any event, these results certainly do not favor the view that the products of gastric digestion are absorbed as soon as they are formed. It is, no doubt, quite different in the intestine, but in the stomach, where pepsin-proteolysis occurs, we have, I think, no grounds for assuming that either peptones or proteoses are rapidly absorbed. Hence, it might perhaps be considered that the results of pepsin-proteolysis in the living stomach are much the same as those obtained in artificial digestion experiments.

Still, there are other differences between natural digestion and artificial proteolysis than those connected with the possible absorption of the more diffusible products of digestion. Thus, in the living stomach there is an ever-increasing secretion of hydrochloric acid, and perhaps also of pepsin, more or less proportional to the extent of proteolysis. On this point Brandl's experiments again give us some light. Thus, it was found that the introduction of an aqueous solution of peptone into the empty stomach led to the secretion of an acid fluid containing on an average 0.24 per cent. HCl, while, under similar conditions, the introduction of sugar or potassium iodide was followed by the secretion of a fluid containing on an average only 0.13 per cent. HCl. Further, the absolute amount of acid found after the introduction of peptone was far greater than when sugar or iodide was introduced, since peptone led to an increase of at least fifty per cent. in the volume of fluid secreted. Hence, proteolysis in the living stomach may give rise to such an increased production and secretion of hydrochloric acid that formation of the terminal products of gastric digestion may be greatly accelerated. That such in fact is the case, I have no manner of doubt, but that it may result in the complete conversion of the so-called primary and secondary proteoses into peptone I very much question. In fact, such examinations as I have made of the stomach-contents after a suitable test-meal have always resulted in the finding of a relatively large amount of proteoses. To be sure, true peptone may be detected and in fairly large amounts, but whenever a quantitative determination of the relative proportion of the two has been made, the proteoses have always been in excess. I have already reported elsewhere the results of some experiments in this direction made on a healthy young man, where the stomach-contents were withdrawn at varying periods after the ingestion of weighed amounts of coagulated egg-albumin. Thus, in one experiment[1] the stomach was thoroughly rinsed with water, after which 138 grammes of finely divided coagulated-albumin, equal to 16 grammes of dry albumin, were ingested. Three-quarters of an hour thereafter, the stomach-contents were withdrawn by lavage and analyzed. As a result, 1.41 grammes of albumoses were separated and weighed, and 0.84 gramme of peptones, the relative proportion being expressed by sixty-two per cent. of albumoses and thirty-seven per cent. of peptones, calculated on the 2.25 grammes of soluble products recovered. This expresses the general character of the results obtained in experiments of this nature, and in my opinion adds emphasis to the statement already made, that complete peptonization is not a feature of pepsin-proteolysis, either in the artificial or in the natural process as it takes place in the living stomach.

Gastric digestion is to be considered rather as a preliminary step in proteolysis, preparatory to the more profound changes characteristic of pancreatic digestion, in which the ferment trypsin is the important factor. We can thus see

[1] Journal of Physiology, vol. xiv., p. 501.

how, as in the case of Czerny's dogs, an animal may be perfectly nourished without a stomach, digestive proteolysis
being carried on solely by the pancreatic fluid. You will
remember that two of the dogs operated on by Czerny and
his pupils lived between four and five years after the
operation, with the stomach completely removed, and yet
during this period they were well nourished and ate all
varieties of food with apparently a normal appetite.[1]
Evidently, then, in some cases at least, digestive proteolysis can be carried on without this preliminary action of
the gastric juice. Ogata[2] arrived at essentially the same
conclusion by the establishment of a duodenal fistula, shut-
ting off the stomach from the intestine by means of a small
rubber ball which could be inflated with water. On then
introducing coagulated egg-albumin and other forms of
proteid matter into the duodenum, he found that digestion
was at least sufficiently complete to satisfy all the demands
of the system. The only unsatisfactory result was with
collagenous foods, which plainly showed the need of a preliminary acid digestion. More recently still, Cawallo and
Pachon,[3] working in Richet's laboratory, have studied the
digestibility of different kinds of proteid foods in a dog,
upon which they had performed a gastrectomy; the
entire fundus, as well as the pyloric portion, of the stomach
having been removed. In an animal so operated upon,
after recovery was complete, solid food, as meat, was completely digested when taken in small quantities at a time.
Raw meat, however, was less completely utilized, the fæces
showing portions of undigested fibres. Still, it was
apparent that intestinal digestion alone was capable of

[1] Bunge's Physiologische und Pathologische Chemie, p. 152.
[2] Ueber die Verdauung nach der Ausschaltung des Magens. Du Bois-
Reymond's Archiv f. Physiol, 1883, p. 89.
[3] Une observation de chien sans estomac. Comptes Rendus hebd. de
la Société de Biologie, December 1, 1893.

accomplishing all that was necessary for the complete nourishment of the animal, when it had once become accustomed to the changed condition of its alimentary tract.

These facts are cited not to belittle the importance of gastric digestion in the nutrition of the body, but rather to emphasize the probability that pepsin-proteolysis is simply a preliminary step in digestion ; that its function is not in the direction of a complete peptonization of the proteid foods ingested, but that its action is especially directed to the production of soluble products, proteoses, which can be further digested in the small intestine, or perhaps directly absorbed after they have passed through the pylorus, or even from the stomach itself to a certain extent.

SOME PHYSIOLOGICAL PROPERTIES OF PROTEOSES AND PEPTONES.

It is very evident from what has been said that all forms of proteid matter, *i.e.*, all the members of the three main groups spoken of in our classification of the proteids, excepting only nuclein, recticulin, and the keratins, are capable of undergoing proteolysis with pepsin-hydrochloric acid. Further, in every case the main products of the transformation are proteoses ; viz., albumoses, caseoses, gelatoses, vitelloses, myosinoses, etc., according to the nature of the proteid undergoing proteolysis ; true peptones being formed in less abundance. Corresponding to each of these groups are primary and secondary proteoses, all possessed of many points in common, both chemical and physiological, yet differing from each other in many minor respects. These are the important products of gastric digestion, of pepsin-proteolysis, and it may be well

to consider for a moment some of the physiological properties of the proteoses and of peptones as well, in order that
we may the better comprehend the general nature of these
substances with reference to their possible action in the
economy.

As far back as 1880, Schmidt–Mülheim[1] discovered that
the injection of aqueous solutions of peptone into the
blood-vessels of living dogs was attended by a series of
remarkable phenomena. Thus, the animal passed at once
into a condition of narcosis resembling that produced by
chloroform, accompanied by a fall of general blood-
pressure so great that the animal was liable to die, as from
asphyxia. Further, there was evidence of some marked
change in the condition of the blood, as indicated by loss
of the power of spontaneous coagulation, while the peptone
itself evidently underwent some alteration, or else was
rapidly eliminated, since it could not be detected in the
blood a short time after its introduction. These experiments, however, were not conducted with true peptone
but with Witte's "peptonum siccum," which at that time,
at least, was composed in great part of proteoses. The
general character of these interesting results was confirmed
by Fano,[2] who found that the injection of so-called peptone in the proportion of 0.3 gramme per kilo. of body-
weight was sufficient to bring about complete narcosis,
together with loss of coagulability on the part of the blood.
Very suggestive, however, was the fact that Fano, on try-
ing similar experiments with the peptone formed by pancreatic digestion, viz., with antipeptone, which presumably
contained a far smaller proportion of proteoses, failed to
obtain like results; the tryptone, so-called, being exceed-

[1] Beiträge zur Kenntniss des Peptons und seiner physiologischen
Bedeutung. Du Bois-Reymond's Archiv f. Physiol., 1880, p. 33.
[2] Das Verhalten des Peptons und Tryptons gegen Blut und Lymphe.
Ibid., 1881, p. 277.

ingly irregular in its action, in many cases producing no effect whatever. The discovery at this date of the several albumoses, and their presence in large amounts in all so-called peptones, led to a study of their physiological action with special reference to the observations of Schmidt-Mülheim and Fano. Politzer,[1] working under Kühne's guidance, was the first to experiment in this direction, and his results are full of interest as throwing light on the action of the individual albumoses. Thus proto, hetero, and deutero-albumose are all active physiologically, giving rise when injected into the veins of dogs and cats to strong narcotic action, varying somewhat in intensity in different individuals. There is also produced a marked fall in blood-pressure, due apparently to vaso-motor paralysis, the action being manifested chiefly, if not wholly, on the splanchnic region. Thus, after an injection of one of these albumoses, the mesenteric vessels are always strongly congested, accompanied frequently by the appearance of a bloody serum in the peritoneal cavity. Narcotic action is manifested only so long as the blood-pressure remains sub-normal, and is due presumably to this marked accumulation of blood in the large abdominal veins, thus leading to anæmia of the brain. Albumoses and peptones injected into the jugular vein likewise produce fever, presumably through some action on the nervous system by which the equilibrium of tissue-metamorphosis is interfered with.[2]

Further, Politzer found that all of the albumoses either delayed or prevented altogether the coagulation of the blood, in conformity with the observations of Schmidt-Mülheim and Fano. In all of these actions the primary

[1] On the Physiological Action of Peptones and Albumoses. Journal of Physiol., vol. 7, p. 283.
[2] Ott and Collmar: Pyrexial Agents, Albumose, Peptone, and Neurin. Journal of Physiol., vol. 8, p. 218.

albumoses appeared most effective, deuteroalbumose least so. Heteroalbumose, however, was constantly most active, especially in delaying the coagulation of the blood. With amphopeptone, there was far less narcosis and less diminution of blood-pressure, while the effect on the coagulability of the blood was more or less variable, frequently being entirely negative. Antipeptone, on the other hand, was found almost wholly wanting in any constant effects, although in one instance deep narcosis was produced. Thus, from Politzer's experiments, it was made clear that the albumoses, when introduced directly into the blood-current, possess a far greater toxic action than either amphopeptone or antipeptone. Albumoses, in sufficiently large doses, were invariably fatal, while peptones never produced fatal results so long as the kidneys of the animal remained intact. The extreme solubility and diffusibility of peptones, coupled perhaps with their marked diuretic action, lead to rapid elimination through the kidneys, and their consequent removal from the system.

Many of these observations made with the albumoses I have repeated with several of the proteoses and peptones more recently studied, as protocaseose, protoelastose, the globuloses, and others. The results may be taken as practically confirmatory of the older observations, and I make mention of them in this general way simply to emphasize the fact that all of the proteoses, though perhaps showing individual peculiarities, are possessed of marked physiological properties, which plainly testify to their toxic nature, when introduced directly into the blood-current.

Young animals are particularly sensitive to the injection of proteoses into the blood, even when the introduction takes place very gradually.[1] Thus, a young, healthy dog

[1] Neumeister: Ueber die Einführung der Albumosen und Peptone in den Organismus. Zeitschr. f. Biol., Band 24, p. 284.

of 2 kilos. body-weight, eight weeks old, died in one hour after the injection into the jugular vein of 1 gramme of protoalbumose-in 20 c.c. of water, thus affording a good illustration of the extreme toxicity of this albumose when introduced directly into the blood.

Of greater interest, physiologically, are the changes the individual proteoses undergo after their injection into the blood. As already stated, peptone so injected may appear in the urine wholly unaltered. Thus, Neumeister[1] has made injections of both amphopeptone and antipeptone in the case of dogs, and was able to detect the peptone very quickly in the urine. I have made like experiments with other forms of peptone and obtained similar results; thus, a pure amphopeptone formed from casein by pepsin-proteolysis (2 grammes in 15 c.c. water) was injected into the jugular vein of a dog weighing 5 kilos. The urine collected during several hours after the injection was heated to boiling, and saturated while hot with ammonium sulphate. The filtrate, on being tested with cupric sulphate and potassium hydroxide, gave a fairly strong biuret reaction for peptone. Another similar experiment made with antipeptone, formed from the myosin of muscle-tissue, gave like results.

With proteoses, however, different results are obtained, as Neumeister[2] first pointed out. These bodies introduced into the blood undergo more or less of a change prior to their excretion in the urine, the change partaking of the character of a hydrolytic cleavage in which the primary proteoses are transformed into secondary proteoses, while deuteroproteoses are changed into peptones. This is not necessarily to be interpreted as meaning that the full equivalent of the proteose injected appears in the urine, but that the portion which is eliminated through the kid-

---

[1] Zeitschr. f. Biol., Band 24, p. 287.   [2] Ibid., p. 284.

neys tends to undergo a transformation somewhere *en route*, akin to the change produced in pepsin-proteolysis. .
As to how common or complete this transformation is under the above circumstances, we have no positive knowledge. Such a hydrolytic change certainly occurs in the case of the dog, and the experimental evidence is in favor of the view that the transformation is effected in the kidneys by the pepsin secreted through the urinary tubules, where there is momentarily a formation of free acid. In the rabbit, on the other hand, no such change occurs; the urine from this animal contains practically no pepsin, and consequently the proteoses eliminated through the kidneys are excreted unaltered. As, however, the experiments of Stadelmann[1] and others have shown that the urine of all carnivora, and of man as well, contains a ferment which, on the addition of a suitable amount of hydrochloric acid, will digest fibrin with formation of the ordinary products of pepsin-proteolysis, it is to be presumed that all proteoses passing through the kidneys will undergo at least some change prior to their excretion in the urine.

However this may be, it is very evident that the proteoses formed in gastric digestion cannot be absorbed as such directly into the blood-current. Introduced into the blood, they behave in such a manner as to warrant the conclusion that they are truly foreign substances, and the system makes a brave endeavor to remove them as speedily as possible. The same may be said of amphopeptones, from which we may conclude that all of these products of pepsin-proteolysis undergo some transformation during the process of absorption, by which their toxicity is destroyed and their nutritive qualities rendered fully available for the needs of the body. Discussion of this question, however, will be left until the next lecture.

[1] Untersuchungen über den Pepsin-fermentgehalt des normalen und pathologischen Harnes. Zeitschr. f. Biol., Band 25, p. 208.

In view of these pronounced physiological properties of the proteoses, it is interesting to recall the now well-known fact that many of the chemical poisons produced by bacteria are proteose-like bodies, chemically, at least, closely akin to the proteoses resulting from pepsin-proteolysis. Thus, Wooldridge[1] as early as 1888 pointed out that an alkaline solution of tissue-fibrinogen exposed to the action of anthrax-bacilli suffered some change, so that when introduced into the blood it possessed the power of producing immunity to anthrax. This observation was verified by Hankin,[2] who further showed that the substance formed by the anthrax-bacilli was a veritable albumose, and that it truly possessed the power of producing immunity. Sidney Martin[3] carried the matter still further, and by growing the anthrax-bacilli in a pure solution of alkali-albuminate prepared from blood-serum, proved the formation of both primary and secondary albumoses, as well as of peptone, leucin, tyrosin, and a peculiar alkaloidal substance of pronounced toxic properties. Martin finds that the albumoses are not as poisonous as the alkaloid, and surmises that the alkaloid is contained in the albumose molecule in the nascent state; further, he suggests that the albumoses in small doses may exert some protective influence, while in larger doses they act as vigorous poisons. How true this may be I cannot say, but my own experience convinces me that the anthrax-bacilli grown in a culture medium composed of alkali-albuminate, prepared from egg-albumin, to which the necessary inorganic salts and some glycerin have been added, do give rise to albumoses and peptones which are truly endowed with toxic properties.

Albumose-like bodies have also been obtained by Brieger

[1] Versuche über Schutzimpfung auf chemischem Wege. Du Bois-Reymond's Archiv f. Physiol., 1888, p. 527.
[2] British Med. Journal, October, 1889.
[3] Proceed. Royal Society, 1890, vol. 48, p. 78.

and Fränkel[1] with the bacillus of diphtheria. These, too, were endowed with powerful poisonous properties, and when introduced into the tissues of the body gave rise to reactions resembling those produced by the Löffler bacillus. In my own laboratory, recent experiments made with the bacillus of glanders have shown that when grown in a slightly acid medium containing alkali-albuminate, albumoses, peptones, and crystalline bodies such as leucin and tyrosin are formed in considerable quantities. Kresling[2] has reported similar results. With the tubercle-bacilli, many like results have been recorded. Thus, among others, Crookshank and Herroun[3] have reported the finding of albumoses, peptone, and a ptomaine when the bacilli have been grown in glycerin agar-agar, and also in fluid media.

Koch[4] has made a special study of the albumose which he considers as the specific toxic agent of the so-called tuberculin. This albumose was found by Brieger and Proskauer[5] to have a somewhat peculiar composition, inasmuch as it contains forty-seven to forty-eight per cent. of carbon and only 14.73 per cent. of nitrogen, agreeing, however, in this respect very closely with the peptone formed from egg-albumin by the action of bromelin.[6] Still more recently, Kühne[7] has made a thorough study of this albumose, as well as of the other products elaborated by the growth of the tubercle-bacillus. He designates all of the peculiar albumoses formed by these bacilli as *acrooalbumoses*. They are endowed with marked chemical and

[1] Untersuchungen über Bacteriengifte. Berlin. klin. Wochenschrift, 1890, p. 241 and 268.
[2] Ueber die Bereitung des Malleins und seine Bestandtheile. Abstract in Jahresbericht f. Thierchemie, Band 22, p. 634.
[3] Journal of Physiology, vol. 12, p. 9.
[4] Deutsche med. Wochenscrift, 1891, p. 1180.
[5] *Ibid.*
[6] Chittenden: On the Proteolytic Action of Bromelin, the Ferment of Pineapple Juice. Journal of Physiology, vol. 13, p. 303.
[7] Weitere Untersuchungen über die Proteine des Tuberculins. Zeitschr. f. Biol., Band 30, p. 221.

physiological properties, causing a rise of temperature when injected into the blood, as well as other phenomena more or less pronounced. It is thus evident there is ample ground for the statement that all nutritive media in which pathogenic bacteria have been planted are liable to contain, sooner or later, toxic substances, many of which at least are closely related to, if not identical with, the albumoses. It is not my purpose, however, to consider these points in detail, nor to quote the many results obtained by other workers in this direction.

I wish merely to call attention to the fact that the proteoses, and likewise the peptones formed by pepsin-proteolysis, are more or less toxic when introduced directly into the blood, and that they share this property with the proteoses formed by bacterial organisms, or by the enzymes which they give rise to. In other words, these primary cleavage or alteration products of the proteid molecule, however produced, are more or less poisonous, and if introduced into the blood-current without undergoing previous change may show marked physiological action. It is, of course, not to be understood that these bodies are all alike. They are surely closely related and possess many points in common, especially so far as their chemical properties are concerned, but their chemical constitution and their physiological action must vary more or less with their mode of origin.

In any event, it is very evident that the proteoses and peptones formed in the alimentary tract by pepsin-proteolysis must undergo some transformation, before reaching the blood-current, by which their peculiar physiological properties are modified. This modification may be associated with a conversion into the serum-albumin, or globulin of the blood. However this may be, the fact remains that these proteoses formed so abundantly during digestion can be absorbed and serve as nutriment for the animal body, but between their formation as a result of proteolysis and their passage into the blood they are exposed to some agency, or agencies, doubtless in the very act of absorption, by which a further transformation is accomplished. With this point we shall be able to deal more in detail in the next lecture.

LECTURE III.

PROTEOLYSIS BY TRYPSIN.

IN pancreatic digestion, proteids are exposed to the action of an enzyme of much greater power than pepsin, one endowed with a far greater range of activity, and couse-quently proteolysis as it occurs in the small intestine becomes a broader and more complicated process. As you well know, the ferment trypsin manifests its power not only in a more rapid transformation of insoluble proteids into soluble and diffusible products, but there is a diversity in the character of the many products formed which testi-fies to the profound alterations this ferment is capable of producing. The primary and secondary products of pepsin-proteolysis, as well as unaltered proteids, are alike subject to these changes, and bodies of the simplest constitution may result in both cases from the series of hydrolytic changes set in motion by this proteolytic enzyme. The power of the ferment as a contact agent is astonishing, for in the case of trypsin no accessory body is necessary to bring out its latent power. Water, proteid, and the enzyme at the body-temperature are all that is necessary to call forth prompt and energetic hydrolytic action.

Moreover, hydrolysis does not stop with the mere pro-duction of soluble proteoses and peptones, but the hemi-portion of the latter is quickly broken down into crystalline bodies, such as leucin, tyrosin, lysin, lysatin, etc. This special characteristic of the ferment testifies in no uncer-tain manner to the existence of inherent qualities in the inner structure of the enzyme peculiar to the body itself. In general properties and reactions, pepsin and trypsin may be closely related; both are products of the katabolic action of specific protoplasmic cells, but the inner nature or structure of the two must be quite different. Pepsin, as we have seen, is powerless to produce any change in proteid bodies unless acids are present to lend their aid. Furthermore, pepsin is limited in its action to the production of proteoses and peptones, while trypsin gives rise to a series of hydrolytic cleavages which result in the ultimate formation of comparatively simple bodies.

Trypsin, however, in its natural environment is dissolved in an alkaline medium. Its proteolytic action is therefore carried on, under normal circumstances, in an alkaline-reacting fluid containing 0.5 to 1 per cent. sodium carbonate, and the proteolytic power of the ferment is unquestionably manifested to the best advantage in such a medium. At the same time, it will act, and act vigorously, in a neutral fluid, and likewise in a fluid having a weak acid reaction, provided there is little or no free acid present. Thus, in experiments[1] on blood-fibrin it was found that, while a solution of trypsin containing 0.5 per cent. sodium carbonate, digested or dissolved 89 per cent. of the proteid in three to four hours at 40° C., a perfectly neutral solution of the ferment, otherwise under exactly the same conditions, digested 76 per cent., and a 0.1 per cent. salicylic acid-solution of the enzyme converted 43 per cent. of the proteid into soluble products.

With hydrochloric acid, trypsin is quickly destroyed, unless there is a large excess of proteid matter present,[2]

[1] Chittenden and Cummins : Studies in Physiol. Chem., Yale University, vol. i., p. 135.
[2] Mays : Untersuchungen aus d. physiol. Institute d. Universität Heidelberg, Band iii., p. 378; also Langley : On the Destruction of Ferments in the Alimentary Canal, Journal of Physiology, vol. iii., p. 263.

which obviously means that the acid in such case exists wholly as combined acid. Indeed, experiments made in my laboratory have shown that as soon as free acid, especially hydrochloric acid, is present in a solution containing trypsin, then proteolytic action is at once stopped. When, however, acids, especially organic acids, are present in a digestive mixture containing an excess of proteid matter, so that the solution contains no free acid (or better, with the proteid matter only partially saturated with acid) then trypsin will continue to manifest its peculiar proteolytic power, although to a considerably lessened extent. Hence, it is evident that the ferment may exert its digestive power under the three possible sets of conditions which, under varying circumstances, frequently prevail in the small intestine.

In considering the general phenomena of proteolysis by trypsin, one is especially impressed by the large and rapid formation of peptone which almost invariably results from the action of a moderately strong solution of the ferment, on nearly every form of proteid matter. To be sure, primary products are first formed, but these are quickly converted into peptone, and a little experience in studying the action of pepsin and trypsin soon reveals the fact that the latter is especially a peptone-forming ferment. In other words, it is peculiarly adapted to take up the work where it has been left by pepsin and, if necessary, carry forward the hydrolytic change even to the extent of a conversion of the entire hemi-moiety into crystalline products.

The primary products of trypsin-proteolysis, however, are not exactly identical with those formed by pepsin. Thus, protoproteoses and heteroproteoses seldom appear in an alkaline trypsin digestion; the proteid matter being in most cases, at least, directly converted into soluble deuteroproteoses,[1] which are then transformed by the further

[1] R. Neumeister: zur Kenntniss der Albumosen. Zeitschr. f. Biol. Band 23, p. 378.

action of the ferment into peptones and other products. Hence, we may express the order of events in the trypsin digestion of a native proteid as follows :

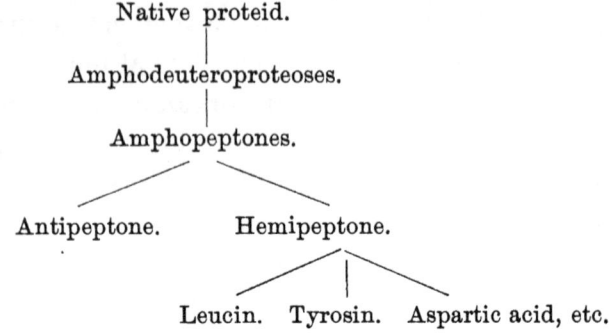

Native proteid.

Amphodeuteroproteoses.

Amphopeptones.

Antipeptone.   Hemipeptone.

Leucin.   Tyrosin.   Aspartic acid, etc.

In the digestion of fresh blood-fibrin with trypsin, there is plainly a preliminary solution of the proteid without any marked transformation or cleavage occurring, the soluble product being apparently a globulin, coagulating at about 75° C.,[1] viz., at approximately the same temperature as serum-globulin. This body, however, quickly disappears, giving place to true deuteroproteoses as the ferment-action commences; for it is not probable that this globulin is a product of enzyme-action, but rather represents a simple solution of the fibrin by the alkaline fluid and salts. In any event, this globulin-like substance is not formed in the pancreatic digestion of coagulated-albumin, serum-albumin, or vitellin, and hence cannot be considered as a true product of trypsin-proteolysis.

The fact that deuteroproteoses are the primary products of trypsin-digestion again emphasizes the natural adaptability of this ferment to the part it has to play in the digestive process. Its natural function is to take up the work where left by pepsin, and carry it forward to the

[1] Jac. G. Otto : Beiträge zur Kenntniss der Umwandlung von Eiweiss-stoffen durch Pancreas-ferment. Zeitschrift f. physiol. Chem., Band 8, p. 129.

necessary point; and hence, when acting upon a native proteid the primary products of its action correspond to the secondary products of pepsin-proteolysis. Trypsin is thus equally efficient in the digestion of all native proteids, but the products of such action are always deutero-proteoses, peptones, and crystalline amido-acids. It is to be remembered, however, that in trypsin-proteolysis the deuteroproteoses and the amphopeptones must necessarily be represented by bodies in which there is a preponderance of anti-groups. In pepsin-proteolysis, as we have seen, the hemi- and anti-groups of the proteid molecule remain more or less united, but in pancreatic digestion, the formation of amphopeptone is quickly followed by the breaking down of a portion of the hemipeptone into leucin, tyrosin, etc. thus leaving a larger proportion of the anti-moiety in the remaining amphopeptone.

Theoretically, at least, in the case of a vigorous and long-continued pancreatic digestion, all of the hemipeptone formed from any native proteid can be converted into crystalline and other products, thus leaving a true anti-peptone resistant to the further action of trypsin. Hence, we are prone to speak of the peptone of pancreatic digestion as antipeptone, although, as can be readily seen, the exact nature of the peptone, *i. e.*, the relative proportion of hemi- and anti-groups it contains, will obviously depend upon the length of the digestion and the strength of the ferment. Again, it is possible, as certain facts seem to suggest, that the amido-acids which are so readily formed from hemipeptone may come in part directly from the hydration of a portion of the hemideuteroproteose, without passing through the preliminary stage of hemipeptone. If so, we have another source of variation in the relative proportion of hemi- and anti-moieties in the deuteroproteoses and peptones of pancreatic digestion. Still again, it is to be remembered that in normal digestive proteolysis, as it occurs in the living

8

intestinal tract, the proteid matter to be acted upon has
already passed through certain preliminary stages in its
transit through the stomach, as a result of which still
further variations in the proportion of hemi- and anti-
groups may be possible.

It is thus plainly evident, in view of the ready cleavage
of the hemi-group into amido-acids, that the primary products of trypsin-proteolysis, the proteoses and peptones, must
necessarily be composed in great part of those complex and
semi-resistant atoms which we include under the head of
the anti-group.   However much one may be skeptical
about the real existence of so-called hemi- and anti-groups,
there is no gainsaying the fact that a given weight of
native proteid, like egg-albumin or blood-fibrin, cannot be
converted wholly into crystalline or other simple products
by trypsin; indeed, it is quite significant that at the end
of a long-continued treatment with an alkaline solution of
the pancreatic ferment, there is usually found about fifty
per cent. of peptone, while the other fifty per cent. of
the proteid is represented mainly by more soluble products, such as the amido-acids.   It is also significant that the
peptone obtained from an artificial pancreatic digestion,
where the proteolytic action has been long-continued and
vigorous, resists the further action of the ferment.   In
other words, it is the so-called antipeptone.   In line with
this result is the fact that the peptones formed in pepsin-
proteolysis, when treated with an alkaline solution of trypsin, are converted into amido-acids and other. bodies of
simple constitution to the extent of about fifty per cent.
This is easily explainable on the ground that the hemi-
portions of the above peptones are broken down into simple products, while the anti-portions remain unchanged,
being resistant to the ferment and thus leading to a separation of the two groups, or at least to the isolation of the
anti-molecules.

There is much that might be cited in further support of these views, but doubtless I have said enough to make it plainly evident that in the pancreatic digestion of any native proteid, not more than one-half can at the most be transformed into crystalline products, while the other half will be represented mainly by a peptone incapable of further change by trypsin. Similarly, the products of pepsin-proteolysis exposed to the action of trypsin may undergo a like separation, the hemi-groups only breaking down into simple products. Hence, the whole theory of the hemi- and anti-moieties of the proteid molecule means simply that of the many complex atoms composing the molecule, one-half are easily decomposable by the pancreatic ferment, while the other half are more resistant and make up the so-called anti-group.

In any active pancreatic digestion of either a native proteid, or of the products of pepsin-proteolysis, the anti-group is represented mainly by antipeptone, although there is often found a small amount of a peculiar antialbumid-like body, insoluble in the weak alkaline fluid. Antipeptones, thus far studied, when entirely free from proteoses, are characterized by a low content of carbon, like the amphopeptones from pepsin-proteolysis. The following table shows the composition of a few typical examples:

COMPOSITION OF ANTIPEPTONES.

	From blood-fibrin.[1]	From blood-fibrin.[2]	From antialbumose.[3]	From casein.[4]	From myosin.[5]
C	47.30	49.59	48.94	49.94	49.26
H	6.73	6.92	6.65	6.51	6.87
N	16.83	15.79	15.89	16.30	16.62
S	0.73	—	—	0.68	1.16
O	28.41	—	—	26.57	26.09

[1] Kühne and Chittenden : Studies in Physiol. Chem. Yale Univer., vol. ii., p. 40.
[2] J. Otto : Zeitschr. f. physiol. Chem., Band 8, p. 146.
[3] Kühne and Chittenden : Zeitschr. f. Biol., Band 19, p. 196.
[4] Chittenden : Studies in Physiol. Chem. Yale Univer., vol. iii., p. 101.
[5] Chittenden and Goodwin : Journal of Physiol., vol. xii., p. 34.

From these data it is evident that, while each individual peptone may have a composition peculiar to itself, they are all alike in possessing a relatively low content of carbon. The antialbumid, however, split off in these hydrolytic changes, like the antialbumid formed by the action of dilute acids at 100° C., is characterized by a correspondingly high content of carbon and a low content of nitrogen. As an illustration, may be mentioned the myosin-antialbumid formed in the digestion of myosin from muscle-tissue by an alkaline trypsin-solution. This body contains 57.48 per cent. of carbon, 7.67 per cent. of hydrogen, 13.94 per cent. of nitrogen, 1.32 per cent. of sulphur, and 19.59 per cent. of oxygen.[1] It is only necessary to compare these figures with those expressive of the compositiou of myosin-antipeptone, to appreciate how wide a gap there is between these two products of trypsin-proteolysis, and both members of the anti-group. Antialbumid, however, is a peculiar product, one which is liable to crop out somewhat unexpectedly, and with varying shades of resistance toward the proteolytic ferments. As formed in pepsin-proteolysis, it is more or less readily soluble in sodium carbonate, and in part readily convertible into antipeptone by trypsin. Still, the same substance, or at least a closely related body, makes its appearance in the form of an insoluble residue whenever a native proteid is digested by trypsin. At times, the amount of this insoluble product may be quite large, even reaching to one-fourth of the total proteid matter;[2] but when so formed in the intestine it must entail a heavy loss of nutriment, for whenever the anti-group is split off after this fashion it becomes very resistant to the further action of the ferment. Separating in this manner from an artificial digestive

[1] Chittenden and Goodwin : Journal of Physiol., vol. xii., p. 36.
[2] Kühne und Chittenden : Ueber die nächsten Spaltungsproducte der Eiweisskörper. Zeitschr. f. Biol., Band 19, p. 196.

mixture, it may be dissolved in dilute caustic alkali, repre-
cipitated by neutralization, and then once again brought
into solution with dilute sodium carbonate.   In this form,
it will yield some antipeptone by the further action of
trypsin, although even then a large amount of the antialbumid is prone to separate out as a gelatinous coagulum,
more or less resistant to the further action of the ferment.

The peculiar action of trypsin, however, as a proteolytic
enzyme is shown in the production of a row of crystalline
nitrogenous bodies of simple constitution whenever the ferment is allowed to continue its action for any length of
time, either on native proteids or on proteolytic products
containing the hemi-group.   This, to be sure, is a fact long
known, but it gains added significance as year by year new
bodies are discovered as products of trypsin-proteolysis
with various forms of proteid matter.   The very character
of the bodies originating in this manner gives evidence of
the far-reaching decompositions involved; decompositions
which are perhaps attributable as much to the innate tendencies of the proteid material as to the specific action
of the ferment.   As representatives of this peculiar line
of cleavage, we have first the well-known bodies, leucin and
tyrosin ; leucin, a body belonging to the fatty acid series,
long known as amido-caproic acid, but now generally considered as amido-isobutylacetic acid, $(CH_3)_2$ CH CH$_2$ CH
(NH_2) COOH ; and tyrosin, a body belonging to the aromatic
group, having the formula $C_6H_4 <^{OH}_{CH_2\ CH\ (NH_2)\ COOH}$,
and known as oxyphenyl-amido-propionic acid.

These two bodies are therefore representatives of two
distinct groups or radicals present in the hemi-portion of
the proteid molecule ; the first belonging to the fatty acid
series, the second to the aromatic group from which come
such well-known bodies as indol, skatol, benzoic acid, and

other substances prominent in proteid metabolism. Moreover, these two hydrolytic products of trypsin-proteolysis
are formed in considerable quantity, at least in an artificial
digestion. Thus, Kühne has reported the finding of 9.1
per cent. of leucin and 3.8 per cent. of tyrosin as the result
of a typical digestion, and I have tried many similar experiments with like results. Further, we know from observations made by different investigators that both leucin and
tyrosin may be formed in considerable quantities in trypsin-
proteolysis as it occurs in the living intestine. But to this
point we shall return later on.

Besides leucin and tyrosin, aspartic acid and glutamic
acid have long been known as decomposition-products of
the vegetable proteids. Thus, both acids were discovered
by Ritthausen and Kreusler[1] in the cleavage of such proteids by boiling dilute acid. Hlasiwetz and Habermann[2]
likewise obtained aspartic acid in large quantity by the
breaking down of animal proteids under the influence of
bromine. Further, Siegfried[3] has recently obtained glutamic
acid as a product of the decomposition of the phospho-
rus-containing proteid, reticulin, from adenoid tissue. As
products of trypsin-proteolysis, Salkowski and Radziejew-
ski[4] found aspartic acid in the digestion of blood-fibrin;
and v. Knieriem[5] likewise obtained it in the digestion of
gluten from wheat. Both of these acids belong to the
fatty acid series, the aspartic acid being a dibasic acid,
$COOH.CH_2$ CH (NH_2). $COOH$, or amido-succinic acid,

---

[1] Verbreitung der Asparaginsäure und Glutaminsäure unter den Zer-
setzungs-producten der Proteinstoffe. Journal f. prakt. Chemie, Band
3, p. 314.
[2] Ueber die Proteinstoffe. Liebig's Annalen, Band 159, p. 304.
[3] Ueber die chemischen Eigenschaften des Reticulirten Gewebes.
Habilitationschrift. Leipzig, 1892.
[4] Bildung von Asparaginsäure by der Pancreas-Verdauung. Bericht.
d. deutsch. chem. Gesellsch., Band 7, p. 1050.
[5] Asparaginsäure, ein Product der künstlichen Verdauung von Kleber
durch die Pancreas-Drüse. Zeitschr. f. Biol., Band 11, p. 198.

while glutamic acid, COOH. C_3 H_5 (NH_2). COOH, is likewise a dibasic acid, known as amido-pyrotartaric acid.

Of more interest physiologically, are the recently discovered nitrogenous bases lysin and lysatinin, or lysatin These two bodies were first identified by Drechsel[1] and his co-workers as products of the decomposition of various proteids, when the latter are boiled with hydrochloric acid and stannous chloride. They were first obtained by Drechsel as cleavage products of casein.[2] Later, Ernst Fischer,[3] working under Drechsel's direction, separated them as decomposition-products of gelatin; while Siegfried[4] obtained them as products of the cleavage of conglutin, gluten-fibrin, hemiprotein, and egg-albumin, by boiling with hydrochloric acid and stannous chloride. In all of these cases it is obvious, from the method of treatment pursued, that the two bodies result from a simple hydrolytic cleavage of the proteid molecule. Hence, it might be assumed that these two bases would likewise be formed in trypsin-proteolysis. This assumption, Hedin,[5] working in Drechsel's laboratory, has proved to be correct, and furthermore he has shown that the amount of these bases formed in pancreatic digestion is not inconsiderable. Thus, as products of the digestion of three kilos. of moist blood-fibrin with an alkaline solution of trypsin, 28 grammes of pure platino-chloride of lysin were obtained, and sufficient lysatinin to establish its identity.

Lysin has the composition of C_6 H_{14} N_2 O_2, being a diamido-caproic acid, a homologue of diamido-valerianic

[1] Der Abbau der Eiweissstoffe. Du Bois-Reymond's Archiv. f. Physiol., p. 248. 1891.
[2] Zur Kenntniss der Spaltungsproducte des Caseins. Ibid., p. 254. 1891.
[3] Ueber neue Spaltungsproducte des Leimes. Ibid., p. 265. 1891.
[4] Zur Kenntniss der Spaltungsproducte der Eiweisskörper. Ibid., p. 270. 1891.
[5] Zur Kenntniss der Producte der tryptischen Verdauung des Fibrins. Ibid., p. 273. 1891.

acid. Hence, this body, like leucin or amido-caproic acid, is a representative of the fatty acid group, the chemical relationship between the two bodies being plainly apparent from their constitution. The constitution of lysatinin is less definitely settled, but apparently it has the composition of a creatin, its formula being $C_6H_{13}N_3O_2$, in which case it might be more appropriately termed lysatin. The special point of interest, however, connected with this latter body as a product of trypsin-proteolysis is the fact that by simple hydrolytic decomposition, all chance of oxidation being excluded, it can break down into urea.[1] For years, chemists have been seeking to trace out the line of cleavage or decomposition by which urea results in proteid metabolism. In the nutritional changes of the body, nearly all the nitrogen of the ingested proteid food is excreted in the fórm of urea, but chemists working with dead food-albumin have been heretofore unable to break down proteid matter directly into urea. This, however, Drechsel has now succeeded in doing, and it is to be especially noted that the line of decomposition or cleavage is simply one of hydration, in which the proteid molecule, either through the action of boiling dilute acids, or through the more subtle influence of the hydrolytic enzyme, trypsin, is gradually broken down into cleavage products, from one or more of which comes lysatin. The very resemblance of this body to creatin suggested that, since the latter breaks down into urea and sarcosin when boiled with baryta water, lysatin might possibly behave in a similar manner. This, as has been previously stated, was found to be the case, and Drechsel obtained from ten grammes of a double salt of lysatin and silver one gramme of urea nitrate, by simple boiling with baryta water.

[1] Drechsel: Ueber die Bildung von Harnstoff aus Eiweiss. Du Bois-Reymond's Archiv f. Physiol., p. 261. 1891.

It is thus evident that a certain amount of urea may come from the more or less direct hydrolysis of proteid matter in the intestinal canal, all but the last steps in the process being the result of the ordinary cleavage processes incidental to trypsin-proteolysis. This fact affords additional evidence of the profound changes set in motion by this proteolytic enzyme. It is not, of course, to be understood that all the urea formed in the body has its origin in this manner. Such a method of decomposition taking place in the intestinal tract would be exceedingly unphysiological and wasteful, but we can readily see how such a line of cleavage might result in inestimable gain to the economy in cases where excess of proteid food has been ingested. Under such circumstances, a portion of the surplus might be broken down directly in the intestine into this urea-antecedent, and thus quickly removed from the system with a minimum amount of effort on the part of the ecomony. Drechsel estimates that about one-ninth of the urea daily excreted may come from the direct decomposition of lysatin, the latter obviously having its origin in trypsin-proteolysis.

Another product of trypsin-proteolysis which has long been recognized, although its real nature has not been known, is tryptophan or proteinochromogen. This body is not only a product of the pancreatic digestion of proteids, but it is also formed whenever native proteids are broken down through any influence whatever, the substance coming presumably from the hemi-moiety of the molecule. It is especially characterized by the bright-colored compound it forms with either chlorine or bromine, so that for a long time it went by the mystical name of the " bromine body." When brought in contact with either of these agents, it immediately combines with them to form a new compound of an intense violet color, termed proteinochrome.

This constitutes the usual test for its presence, a little bromine water, for example, quickly bringing out a violet color when added to a fluid containing the chromogen. The body is readily soluble in alcohol, and hence can be easily separated from the primary products of trypsin-proteolysis, such as the proteoses and peptones. Krukenberg considered the substance not a true proteid, but rather a body belonging to the indigo-group; but Stadelmann, who has given the matter a very thorough investigation, comes to the conclusion that it is truly a proteid body, in part closely related to peptone, although in many ways quite different.

The following composition of bromine proteinochrome, as determined by Stadelmann,[1] shows the general nature of the compound formed when bromine combines with the chromogen :

	A	D
C	49.00	48.12
H	5.28	5.09
N	10.99	11.92
S	3.77	3.10
O	11.01	12.00
Br	19.95	19.77

From the average of the several results obtained, it would appear that the proteinochromogen, which could not be isolated by itself in sufficient purity for analysis, must contain approximately 61.02 per cent. of carbon, 6.89 per cent. of hydrogen, 13.68 per cent. of nitrogen, 4.69 per cent of sulphur, and 13.71 per cent. of oxygen. As a proteid-like body, it is thus especially characterized by an exceedingly high content of carbon and a high content of sulphur. As a product of trypsin-proteolysis, it must presumably come from the cleavage of hemipeptone, which, however, contains only 0.75 per cent. of sulphur. But as

[1] Ueber das beim tiefen Zerfall der Eiweisskörper entstehende Proteinochromogen, den die Bromreaction gebenden Körper. Zeitschr. f. Biol., Band 26, p. 521.

we have seen, this latter body breaks down by further cleavage into substances such as leucin, tyrosin, lysin, etc., which contain-no sulphur whatever, and as there is no elimination of sulphur in this process through formation of hydrogen sulphide gas or otherwise (putrefaction being excluded by the presence of either chloroform or thymol), it follows that this surplus sulphur must accumulate somewhere. The high content of carbon, however, in proteinochromogen is sufficient evidence that the substance cannot have its origin in a simple cleavage of hemipeptone. On the other hand, everything points to a sythetical process, in which two or more cleavage products of the proteid molecule combine and form a new body, such as proteinochromogen, containing all the sulphur cast off from the hemipeptone in the production of the crystalline bodies, and having in itself properties common to peptone and to a body of the indigo-group, the latter obviously coming from some aromatic antecedent.

In view of the apparent complexity of the processes attending trypsin-proteolysis, it is not strange that even simpler substances than those already described should make their appearance. Thus, when it was suggested that ammonia, NH_3, might be formed under the influence of trypsin, it was not considered at all improbable, for in the hydrolytic decomposition of proteids by boiling dilute acid, as well as by baryta water, it had long been known as a prominent product. Obviously, in trypsin-proteolysis, the one thing to be guarded against in proving the formation of ammonia is the contaminating influence of bacteria. Hirschler,[1] however, with a full recognition of this danger, made digestions of blood-fibrin with trypsin extending only through four hours and at a temperature of 32° C., and yet

[1] Bildung von Ammoniak bei der Pancreasverdauung von Fibrin. Zeitschr. f. physiol. Chem., Band 10, p. 302.

he obtained plain evidence of the formation of ammonia. Stadelmann,[1] with still greater precautions to exclude all bacterial agencies, using boiled fibrin as the material to be digested and thymol to prevent any possible infection of the digestive mixture, proved conclusively that ammonia was formed as a result of trypsin-proteolysis. Thus, in the digestion of 35 grammes of boiled blood-fibrin with 60 c. c. of a pancreas infusion for three days, 20.8 milligrammes of NH_3 were developed, presumably coming from the liberation of a certain amount of nitrogen attendant upon the formation of such bodies as leucin and tyrosin, which contain considerably less nitrogen than their direct antecedent hemipeptone, or the original proteid. We thus have striking proof of the ability of this peculiar proteolytic enzyme to set in motion hydrolytic changes which may extend even to the production of such simple substances as ammonia, thus making still more striking the parallelism between trypsin-proteolysis on the one hand, and the artificial hydrolysis produced by boiling dilute acids on the other.

In view of all these facts regarding the nature of the products obtainable by pancreatic proteolysis, it is very evident that many chemical changes may take place side by side in a vigorous pancreatic digestion of proteid matter. We know without a shadow of doubt that all of the bodies enumerated as products of pancreatic digestion are the results of trypsin-proteolysis, and not the products of putrefactive changes. Bacteria, it is true, are able to produce many like products, and in the living intestinal tract exercise an important influence, especially in the breaking down of resistant forms of proteid matter, and in the decomposition of surplus material which has escaped the pancreatic ferment. But all the bodies described above are readily

. [1] Bildung von Ammoniak bei Pancreasverdauung von Fibrin. Zeitschr. f. Biol., Band 24, p. 261.

obtainable by trypsin-proteolysis under conditions which exclude all possibility of bacterial action.

Granting, then, as we must, that these various bodies are all products of pancreatic proteolysis when the process is carried on in beakers or flasks, we need to consider next how far such bodies appear in the natural process as it takes place in the living intestine. We know indeed that the natural and the artificial processes are very much alike so far as the qualitative results are concerned, but what differences there may be between the quantitative relationships in the two cases is less certain. One might naturally reason that, with the facilities for rapid absorption that exist in the small intestine, trypsin-proteolysis would rarely proceed beyond the peptone stage, yet we have ample evidence that, under some circumstances at least, both leucin and tyrosin are formed in considerable quantities in the intestine.

It obviously makes a very great difference to the economy in what form the proteid matter ingested leaves the intestine on its way into the blood-current. It has been more or less generally assumed that, under the ordinary circumstances existent in the intestinal tract, the crystalline and other bodies coming from the more profound changes incidental to trypsin-digestion are rarely formed, mainly on the ground that such transformations would entail great loss of nutritive material to the blood. Years ago, Schmidt-Mülheim [1] made a series of experiments on the changes which proteid foods undergo in different portions of the alimentary tract, from which he concluded that leucin and tyrosin are formed in such small quantities in natural pancreatic digestion that they represent only a very small part of the nitrogen absorbed from the intestine. This conclusion has been more or less generally

[1] Untersuchungen über die Verdauung der Eiweisskörper. Du Bois-Reymond's Archiv f. Physiol., 1879, p. 39.

accepted, especially as several observers have reported finding only small amounts of these bodies in the intestine • under what might be assumed to be favorable circumstances for their formation. In artificial digestions, on the other hand, as we have seen, leucin and tyrosin, together with the other simple bodies described, may appear in large quantities. Obviously, two suggestions present themselves as explanatory of this difference ; either there is such a rapid absorption of these crystalline products from the intestine that they cannot be detected other than as mere traces, or else the natural process takes a different course from the artificial, owing to the rapid withdrawal from the intestine of the antecedent of the leucin and tyrosin, viz., the hemipeptone.

Concerning this point, Lea[1] has recently reported some experimental evidence obtained by a comparative study of artificial pancreatic digestion as carried on in a flask, with similar digestions carried on in parchment dialyzer tubes, the latter so arranged that the diffusible products of proteolysis can pass from the tube into the surrounding fluid. As Lea justly says, this whole question of the formation of leucin by proteolysis is a very important one, since it bears closely upon one of the possible methods by which urea may be quickly formed from proteid food. Thus, we have evidence that when leucin is administered to mammals a portion of its nitrogen, at least, quickly reappears as urea and uric acid in the urine.[2] Further, there is a certain amount of evidence that this transformation takes place in the liver, viz., in the organ where leucin absorbed from the intestine would naturally be first carried.[3]

[1] A Comparative Study of Artificial and Natural Digestions. Journal of Physiology, vol. xi, p. 226.

[2] E. Salkowski: Weitere Beiträge zur Theorie der Harnstoffbildung. Zeitschr. f. Physiol. Chem., Band 4, pp. 55 and 100.

[3] W. Salomon: Ueber die Vertheilung der Ammoniaksalze im thierischen Organismus und über den Ort der Harnstoffbildung. Virchow's Archiv Band 97, p. 149.

Obviously, the main point to be gained in a dialyzer-experiment is the removal of the soluble products of digestion as soon as they are formed; but peptones are not rapidly diffusible, and the process, as noted under the head of gastric digestion, cannot be considered in any sense as yielding the same results as might be obtained in the living intestine.   Still, the method offers a closer approach to the natural process than when carried on in a flask, and the results are of interest.   Thus, Lea finds in the first place that in a dialyzer-digestion the proteid is more quickly dissolved, and that there is far less tendency for the formation of an insoluble antialbumid with its natural resistance to the ferment.   Still, it is to be noticed that the amount of this antialbumid-residue formed by trypsin-proteolysis in a flask is mainly dependent upon the strength of the ferment solution, and the character of the proteid undergoing digestion.   If the latter is in a fairly digestible form, and the enzyme solution reasonably active, then even the flask-digestion may show almost no residue of antialbumid.   Yet there is at least a shade of difference in the two cases, which may be expressed by the statement that trypsin-proteolysis, as carried on in a dialyzer-tube, is prone to give less insoluble antialbumid than a corresponding digestion in a flask.   Further, the amount of leucin and tyrosin formed in a flask-digestion is always greater than in a dialyzer-digestion, other conditions being equal. Naturally, these results help us very little in drawing any conclusions regarding the extent to which leucin and tyrosin may be formed in the intestine.   They merely emphasize the fact that the withdrawal of a certain quantity of hemipeptone from the digestive mixture tends to reduce by so much the yield of leucin and tyrosin.   It is hardly to be assumed, however, that the rate of withdrawal of peptone from the intestine can keep pace with its formation, especially when it is remembered that the proteid matter coming into the small intestine, owing to its preliminary treatment in the stomach, is in a comparatively digestible condition. Further, the pancreatic juice is a remarkably active fluid, and proteolysis under its influence must make rapid strides. I can easily conceive that proteolysis by trypsin, when carried on in a flask, may lead to the formation of much larger amounts of leucin and tyrosin, and of other bodies as well, than occurs in the natural process; but there is certainly no ground for the belief that leucin and tyrosin are wholly wanting in pancreatic proteolysis as it occurs in the intestine.

With a view to obtaining some positive evidence on this point I have tried a few experiments on animals, the results of which have convinced me that, in the case of dogs, at least, both leucin and tyrosin may be formed in natural pancreatic digestion in considerable quantities. Thus, in one experiment a good sized dog, kept without food for two days, was fed four hundred grammes of chopped lean beef at 8 A. M.   At 2 P. M. the animal was killed and the intestine ligatured close to the pylorus.   The lower end of the small intestine was likewise ligatured.   The portion inclosed between the two ligatures was then removed from the body, and the contents of the intestine pressed and rinsed out with distilled water.   In the stomach, was found a small amount of semi-digested matter weighing about fifty grammes.   The material obtained from the intestine was strained through mull, the fluid rendered faintly acid with acetic acid, and heated to boiling.   The clear filtrate from this precipitate was concentrated to a very small volume, and while still hot precipitated with a large amount of ninety-five per cent. alcohol.   A small gummy precipitate resulted, which was thoroughly extracted with boiling alcohol and the washings added to the alcoholic filtrate. The precipitate contained some deuteroproteose and a small amount of true peptone. The alcoholic fluids were evaporated to a small volume and set aside in a cool place. As a result, quite a separation of leucin and tyrosin occurred in the characteristic crystalline forms. No attempt was made to effect a quantitative separation of the two bodies, but the mixed precipitate finally obtained weighed, after recrystallization, over three-fourths of a gramme. Leucin was plainly in excess, but considerable tyrosin must have been left in the alcoholic precipitate, owing to its greater insolubility in this menstruum. This experiment is almost a counterpart of one reported by Lea,[1] and like his indicates that both leucin and tyrosin may be formed in not inconsiderable quantities by pancreatic proteolysis as it occurs in the intestine. This being so, one is naturally called on to explain "the physiological significance of a process which at first sight appears to result in a degradation of the potential energy of proteids, under conditions such that the energy set free can be of little use to the economy."[2] But it is quite possible, as Lea has suggested, that these amido-bodies have an important part to play in some of the synthetical or other processes of the organism, and that their formation is consequently necessary for the well-being of the body. Whether this is so or not, we may well consider the formation of these amido-acids in pancreatic proteolysis as a means of quickly ridding the body of any excess of ingested proteid food, with the least possible expenditure of energy on the part of the system. This has always seemed to me the probable purpose of the profound changes which the pancreatic ferment is capable of inducing.

[1] Journal of Physiology, vol. xi, p. 255.
[2] Lea : loc. cit.

9

The primary object of both gastric and pancreatic prote-. olysis is to render the proteid foods more easily available. for the needs of the economy, viz., to aid in their absorption and consequent distribution to the master tissues and organs of the body. This is doubtless fully accomplished by the formation of the so-called primary and secondary products of proteolysis, i. e., the proteoses and peptones which are, comparatively, not far removed from the mother-proteid, except in solubility and other minor points. In the ferment trypsin, however, we have a special agent endowed with the power of carrying on the hydrolytic cleavage to a point where exceedingly simple bodies result, and through whose agency any excess of proteid material in the intestinal canal may be quickly broken down into a row of products easily removed from the system. It is to be remembered, however, that the very nature of the proteid molecule precludes the possibility of anything like a complete decomposition into crystalline or other simple products. Full fifty per cent. of the peptone formed must be antipeptone, which cannot be further changed by trypsin under any circumstances, so that, whether the amount of proteid in the intestine be large or small, or whether it is exposed for a longer or shorter period to trypsin-proteolysis, there will always be a fairly large amount for absorption. This may well be considered as one of the reasons for the peculiar structure of the proteid molecule, the anti-group being always available for the direct nutrition of the body, while the representatives of the hemi-group, especially when proteid is present in excess, can be quickly and readily broken down into simple products. In other words, the direct formation of these simple bodies in the intestine furnishes a short path to urea, thus leading to the rapid elimination of any excess of proteid material.

We may well attribute to the epithelial cells of the intestine the power, under normal circumstances, of regulating and controlling, even though indirectly, the order of events in the intestine. Just as the so-called secreting cells of the *tubuli uriniferi* may lose for a time their power to pick out from the blood material destined for the urine, being clogged or exhausted by continued effort, so the epithelial cells of the intestine, which play such an important part in the absorption of proteid matters from the alimentary tract, may, in the presence of an excess of proteid matter, become temporarily exhausted, and, refusing passage to the proteoses and peptone formed by proteolysis, render possible further hydrolytic cleavage into leucin, tyrosin, lysatin, etc.; bodies which, by one method or another, can be readily transformed into urea. At the same time, as already stated, it seems more than probable that some formation of these amido-acids always occurs in the intestine, and that these bodies have some specific part to play in the normal processes of metabolism going on in the body. The more one studies the processes of nutrition in general, the more one is impressed with the view that there is a purpose in everything, and that the formation of even such bodies as leucin and tyrosin may be connected with hidden processes, the key to which has not yet been found. We see an analogous case, perhaps, in the action of the inorganic salts in nutrition, some of which, at least, neither undergo change themselves nor induce changes in other substances, and yet we know their presence is indispensable for keeping up the normal rhythm of the nutritional processes of the body.

ABSORPTION OF THE MAIN PRODUCTS OF PROTEOLYSIS.

In ordinary proteolytic action, both in the stomach and intestine, it is very apparent that the primary products of proteolysis, the proteoses and peptones, are the chief products formed, and that under normal circumstances the greater portion of the proteid food finds its way from the alimentary canal into the blood, after transformation into one or more of these two classes of products. At the same time, it must be borne in mind that even the acid-albumin formed by pepsin-hydrochloric acid may be absorbed without undergoing further change. The view once held, that the rate of absorption from the alimentary tract stands in close relation to the diffusibility. of the products formed, and that non-diffusible substances are incapable of absorption, is no longer tenable. Absorption from the intestine is to be considered rather as a process involving the vital activity of the epithelial cells of the intestinal mucous membrane, where chemical affinities and other like factors play an important part in determining the rate and order of transference through the intestinal walls into the blood and lymph. Thus, we have abundant evidence that native proteids which have not undergone proteolysis may be absorbed from the intestine, at least to a certain extent, provided they have been dissolved; *i. e.*, converted into acid-albumin, or alkali-albuminate, by the gastric or pancreatic juice. We have a practical demonstration of this possibility in the early experiments of Voit and Bauer,[1] as well as in many later ones that need not be mentioned here. Further, the recent experiments of Huber[2] have given us quantitative data on the rate of

[1] Ueber die Aufsaugung im Dick und Dünndarm, Zeitschr. f. Biol., Band 5, p. 562.
[2] Ueber den Nährwerth der Eierklystiere, Arch. f. klin. Med., Band 47, p. 495.

absorption of fluid egg-albumin when introduced into the large intestine in the form of a clyster, showing that even fairly large amounts of a natural proteid may be absorbed without undergoing proteolysis if mixed with a neutral salt, like sodium chloride. To be sure, the rate of absorption is greatly increased when the albumin has been peptonized, but still absorption of the native proteid is possible without the agency of proteolytic enzymes. When, however, large amounts of egg-albumin are introduced into the intestine, albuminuria may result, as you very well know.

Moreover, it is well known that the proteids of muscle-tissue, in the form of syntonin, may be absorbed from the large intestine without undergoing further hydration. When introduced into the rectum of a hungry dog, the excretion of urea may be at once increased and the animal brought into a condition of nitrogenous equilibrium; absorption taking place from a portion of the large intestine, where proteolysis is never known to occur.[1]

Again, Neumeister[2] has shown that the direct introduction of syntonin, alkali-albuminate, crystalline phytovitellin, as well as pure serum-albumin, into the blood of the jugular vein is not attended with the appearance of albumin in the urine. On the contrary, the proteid matter so introduced appears to be assimilated and utilized for the needs of the organism. Evidently, then, these substances are not to be considered as foreign bodies, for if so the kidneys would undoubtedly make some effort to remove them from the circulation. It is to be noted, however, that all native proteids are not assimilated in this manner, as casein,[3] gelatin,[4] and especially egg-albumin. Thus,

[1] Eichhorst: Ueber die Resorption der Albuminate im Dickdarm, Pflüger's Archiv f. Physiol., Band 4, p. 570.
[2] Zur Physiologie der Eiweissresorption und zur Lehre von den Peptonen, Zeitschr. f. Biol., Band 27, p. 309.
[3] Neumeister: Sitzungsber. der Physik. med. Gesellsch. zu Würzburg, 1889, p. 73.
[4] F. Klug: Pflüger's Archiv f. Physiol., Band 48, p. 122.

J. C. Lehman,[1] working under Kühne's direction, observed that the injection of a carefully filtered solution of egg-albumin into the veins of a dog was always accompanied by albuminuria, while similar injections of Lieberkühn's sodium albuminate, or of syntonin from frog's muscle, failed to show any such result.

While these observations tend to show that some native proteids may be absorbed from the alimentary tract without previously undergoing proteolysis, it is not to be understood that any considerable quantity is so absorbed under normal circumstances. Doubtless, when small amounts of proteid food are taken, its denaturalization by the primary action of the gastric or pancreatic juice, viz., its conversion into syntonin or alkali-albuminate, may be sufficient to insure its partial absorption, but digestive proteolysis is unquestionably a necessary preliminary to any general absorption, and there can be no manner of doubt that the greater portion of the proteid food is absorbed as proteoses and peptone. Peptones, as we have seen, are possessed of a higher endosmotic equivalent than the proteoses, but we need to keep continually in mind the possibility that the selective power of the epithelial cells of the intestinal mucosa may lead to as rapid transferrence of the proteoses as of the more diffusible peptones. It is not to be understood by this, however, that diffusibility is of no consequence in determining the rate of absorption. Surely, everything else being equal, the more diffusible the substance the more rapid will be its passage from the intestine into the blood-current. The more the process of absorption is studied, however, the more clearly do we see its dependence upon the functional power of the living epithelial cells, a fact which plainly emphasizes the physiological nature of the process.

[1] Virchow's Archiv, Band 30, p. 593.

Further, as already stated, absorption of proteid food-stuffs, or their products, from the alimentary tract, is, under ordinary-circumstances at least, limited to the intestine; from the stomach there is comparatively little absorbed, and if necessary we might advance this fact as an important argument against the theory of general absorption of proteids in the form of acid-albumin. Even such indifferent fluids as water, or physiological salt solution, are absorbed with extreme slowness from the stomach;[1] this organ showing very little ability to take up water even when the blood-vessels are dilated, as after the ingestion of food.

This brings us to a very important point in connection with the utilization by the system of the ordinary products of proteolysis. The latter, as we have seen, are mainly proteoses and peptones, and yet all the evidence points clearly to the fact that these substances are never present, at least in any quantity, in the blood or lymph, even when digestive proteolysis is at its height. Further, the very nature of the proteoses and peptones, their marked physiological action when they are introduced directly into the circulating blood, their rapid excretion, either as proteoses or peptones, by the kidneys when so introduced,[2] all indicate that they are foreign substances, totally out of their natural environment when introduced into the blood-current. And yet we very well know that proteoses especially are possessed of high nutritive qualities; they are abundantly able to support animal life. Thus, Politzer[3] found by feeding experiments with heteroalbumose, dysalbumose, and protoalbumose, that these bodies taken into

[1] J. S. Edkins: The Absorption of Water in the Alimentary Canal. Journal of Physiol., vol. 13, p. 445.
[2] Franz Hofmeister: Ueber das Schicksal des Peptons im Blute, Zeitschr. f. physiol. Chem., Band 5, p. 125.
[3] Ueber den Nährwerth einiger Verdauungsproducte des Eiweisses, Pflüger's Archiv f. Physiol., Band 37, p. 301.

the stomach have the same nutritive value as méat. Vari-. ous feeding experiments with proteoses from different, sources, carried out in my laboratory on young dogs, have shown conclusively that for short periods of time, at least, these hydrolytic cleavage products are fully as capable of sustaining the nitrogenous equilibrium of the body as the proteids from which they are derived. In fact, the results obtained favor the view that the proteoses, weight for weight, possess a higher nutritive value than fresh beef.[1] It may be questionable, however, whether such a result would follow in experiments conducted over longer periods of time, but of this we may be certain, that the proteoses formed in the alimentary tract can be absorbed and utilized by the system without their exerting any toxic action whatever.

Consequently, we are forced to the conclusion that these primary products of proteolysis, so important in the nutrition of the animal body, must undergo some change during the process of absorption, by which they are converted into new bodies, less toxic in their nature, and better adapted for the direct nutritional needs of the organism. The same statement applies likewise to peptones.

The fact that peptones are not discoverable in the blood and lymph, even during or after active digestion, was practically ascertained years ago by such well-known workers as Maly, Adamkiewicz, and others. The natural supposition following this observation was that the products of proteolysis underwent some change in the hepatic cells; but this view was soon shown to be untenable by examination of the portal blood, which was found to be as free from peptone as the blood of the hepatic vein. Neumeister,[2] using the more modern methods of work and with the

---

[1] Compare Hildebrandt, Zur Frage nach dem Nährwerth der Albumosen, Zeitschr. f. physiol. Chem., Band 18, p. 120.

[2] Ueber die Einführrung der Albumosen und Peptone in den Organismus, Zeitschr. f. Biol., Band 24, p. 277.

wider knowledge gained during these latter years, has shown conclusively that proteoses and peptones are never present in the blood, even when these substances are contained in the intestine in fairly large amounts. I can corroborate these statements from the results of my own experiments in this direction. Thus, I have taken a dog in full digestion, fed with an abundance of meat, and collecting the blood from the carotid artery have made a careful examination for peptone, by the following method: The blood was allowed to flow directly into a dilute solution of ammonium sulphate, sufficiently strong to prevent coagulation, and then shaken with ether to rupture the red blood-corpuscles. The solution, freed from ether, was next saturated with crystals of ammonium sulphate, by which the proteid matter was completely precipitated. The clear filtrate was then concentrated somewhat, the excess of the ammonium salt removed by filtration, and the filtrate carefully tested for peptone by addition of a large volume of a saturated solution of potassium hydroxide and a few drops of a dilute solution of cupric sulphate. The test was wholly negative, although the intestine of the animal showed the presence of both peptone and proteoses. This result, as I have said, is simply confirmatory of work done by others in this direction, notably Neumeister, and illustrates the statement that peptones are not to be found in the circulating blood, even after a full proteid diet. In this connection it is to be remembered that we have abundant proof of the rapid disappearance of both proteoses and peptones[1] from the intestine, either by absorption or otherwise. They certainly disappear, and, as we have seen, are not to be found in the blood. Further, Neumeister has confirmed the original observation of Schmidt-

---

[1] Rohmann : Ueber Secretion und Resorption im Dünndarm, Pflüger's Archiv f. Physiol., Band 41, p. 440.

Mülheim,[1] that both chyle and lymph are practically free from proteoses and peptone, thus again forcing us to the conclusion that the primary products of proteolysis must undergo change prior to their passage into the blood or lymph.

Many observations lend favor to the view that a transformation of some kind takes place in the intestine itself, not indeed in the lumen of the tube, but somewhere in the walls, through which the peptones must pass before reaching the blood. Thus, peptones placed in contact with pieces of the isolated, though still living, intestine, after a time completely disappear from view,[2] so completely that no reaction can be obtained even by the most delicate of tests. In support of this statement I may cite the results of several of my own experiments which certainly furnish evidence that true peptones undergo profound alteration by simple contact with the living mucous membrane of the small intestine. The method employed was similar to that made use of some years ago in a study of the influence of peptone on the post-mortem formation of sugar in the liver.[3] A large, well-nourished rabbit was killed by severing the carotid artery and the blood collected and defibrinated. Of this, 50 c. c. were mixed with an equal volume of 0.5 per cent. salt solution containing 1.25 grammes of pure amphopeptone, prepared from egg-albumin, the mixture obviously containing 1.25 per cent. of peptone. The fluid was transferred to a large, roomy flask, provided with a stopper having two holes, in one of which was fitted a

[1] Du Bois-Reymond's Archiv f. Physiol., p. 33, 1880.

[2] Salvioli : Eine neue Methode für die Untersuchung der Functionen des Dünndarms, Du Bois-Reymond's Archiv f. Physiol., 1880. Supplement Band,p. 112. Neumeister : Zur Physiologie der Eiweissresorption und zur Lehre von den Peptonen, Zeitschr. f. Biol., Band, 27, p. 324.

[3] Chittenden and Lambert : Studies in Physiol. Chem., Yale Univer., vol. i., p. 171.

long glass tube reaching below the fluid. The flask, with its contents, was then placed in a suitable water-bath at a temperature of 40° C.

The small intestine of the rabbit was carefully separated from the mesentery and from the pancreatic gland, and the upper portion cut open and quickly washed free from any contained matter or adherent secretions, by repeated immersion in 0.5 per cent. salt solution warmed at 40° C. This was repeated until the tissue was quite free from all impurities, after which it was cut into small pieces and immersed for a moment in a 0.5 per cent. solution of sodium chloride containing 1.25 per cent. of peptone. The tissue was then carefully collected on coarse muslin, allowed to drain, and then quickly transferred to the flask containing the warm blood and peptone. This mixture was kept at 40° C. for two hours, a slow current of air being bubbled through the fluid during the entire period. At the expiration of this time the fluid was separated from the pieces of tissue by filtration through muslin, and then saturated with ammonium sulphate after the usual method for the separation of albumoses, etc. On now testing a portion of the clear filtrate for peptone by the biuret test, not a trace of a reaction could be obtained. The entire amount of proteid matter present was precipitated by the ammonium salt, thus showing that the peptone originally added had been completely transformed into something precipitable by saturation of the fluid with ammonium sulphate. That this transformation of the peptone was accomplished mainly through the action of the intestine, was shown by a parallel experiment, in which all of the above conditions were duplicated, omitting only the pieces of intestine. Here, however, on testing the filtrate from the ammonium sulphate-precipitate, a strong biuret reaction was obtained, thus proving the presence of at least some unaltered peptone.

This experiment is almost a counterpart of one reported - by Neumeister, and like his, testifies to the probability. that the peptones formed in the alimentary tract, as a result of proteolysis, undergo retrogression through the agency of the epithelial cells of the intestinal walls during their absorption. I have tried similar experiments with deutero-proteose, notably with deuterocaseose, and have obtained corresponding results. The same method may be employed as that already outlined, although of course the deutero-caseose is in great part precipitated by saturation with ammonium sulphate. Still, this form of deuteroproteose, β deuterocaseose, as I have elsewhere noted, is very slowly precipitated by the ammonium salt. Consequently, it is an easy matter to demonstrate that this proteose, on treat-ment with the intestinal mucosa in the presence of blood at the body-temperature, is transformed into something completely and readily precipitable by ammonium sul-phate; the filtrate from the latter failing to show any biuret reaction, although the corresponding control experi-ment without the intestine gives a bright violet color with cupric sulphate and potassium hydroxide.

Hence, we are certainly justified in saying that both peptones and proteoses undergo some retrogression when in contact with the walls of the intestine. Moreover, there is some evidence that the proteoses, before under-going such a transformation, are first converted into pep-tone by the action of the intestinal walls, a statement which will apparently apply to both the primary and sec-ondary proteoses. This primary action of the intestinal walls is not considered as due to any adherent trypsin, or to possible traces of succus entericus, but rather as a part of the action of the living epithelial cells, or perhaps as connected with the possible presence of lower organisms not removable from the intestinal wall by ordinary washing.

The transformation of peptones by the substance of the intestine is apparently common to the intestinal tract of many animals, and perhaps to all, and indeed can also be accomplished by the liver.[1] This latter fact is of some importance, since it adds weight to the supposition that this peculiar action of the intestine cannot be due to the possible presence of trypsin; a view which is strengthened by the fact that a glycerin-extract of the intestine has no action on amphopeptone. Certainly, the latter shows no diminution in the strength of the biuret reaction after long contact at body temperature with such an extract. Further, it has been shown that antipeptone, which is not affected by the pancreatic enzyme, suffers the same change as amphopeptone by contact with the intestine. Far more probable is it that retrogression or transformation of peptone by the substance of the intestine, is due to the vital activity of some or all of the epithelial cells of the intestinal mucosa; a characteristic possibly shared by some or all of the hepatic cells of the liver. The kidney-cells certainly do not possess this power, but we can see a special fitness in the liver-cells being endowed with the ability to quickly break down, or transform, any peptone or proteose that might by chance escape unaltered from the intestinal tract. Shore,[2] however, inclines to the view that the hepatic cells do not possess this power to any great extent, in opposition to the older views of Plòsz and Gyergai,[3] as well as of Seegen[4] and of Neumeister.

With reference to the action of the stomach-mucosa on proteoses, it has been shown[5] that when relatively large

[1] Neumeister: Zeitschr. f. Biol., Band 27, p. 332.
[2] On the Fate of Peptone in the Lymphatic System, Journal of Physiol., vol. xi., p. 528.
[3] Ueber Peptone und Ernährung mit denselben, Pflüger's Archiv f. Physiol., Band 10, p. 536.
[4] Zur Umwandlung des Peptons durch die Leber, Ibid., Band 37, p. 325.
[5] Hildebrandt: Zur Frage nach dem Nährwerth der Albumosen, Zeitschr. f. physiol. Chem., Band 18, p. 180.

amounts (5 grammes) are introduced into the stomach of a·
rabbit, the pylorus being ligatured, both proteoses and·
peptones may appear in the urine, thus indicating that
while they may be absorbed to some extent under the
above conditions, the proteoses are not readily transformed
into native proteids without exposure to the intestine.
Smaller amounts (2 grammes), however, may, under the
above conditions, be completely transformed; at least
Hildebrandt claims this to be the case, mainly on the
ground that after the introduction of albumoses into the
stomach, the pylorus being ligatured, no trace of them
can be found in the urine. The same observer also
claims that blood-serum, in the case of dogs, is able to
transform albumoses into ordinary serum-globulin. Certainly, after intra-venous injection, proteoses disappear
from the blood, but, as we shall see later on, a certain
amount, at least, may be transferred to the lymph. It is
also claimed that when albumoses are injected subcutaneously, neither albumoses nor peptones are to be detected in
the urine. This, however, seems hardly probable in the
light of what has been said, and especially in view of the
fact that Neumeister's experiments tend to show that even
0.1 gramme of albumoses introduced subcutaneously may
give rise to temporary albuminuria.

Assuming for the moment that the chief products of
proteolysis, i. e., the proteoses and peptones, are, during
the act of absorption, transformed through the vital
processes of the epithelial cells of the intestine into serum-
albumin, or globulin, and absorbed as such into the blood,
we may well consider whether such transformation, i. e., a
retrogression into a native proteid again, is inconsistent, or
out of harmony, with the general character of the changes
known to occur in the body. In attempting to answer this
question we need not look far to find a perfectly analogous

case. Thus, in the digestion of starchy foods by the amylolytic ferments of both the saliva and the pancreatic juice, the carbohydrate material undergoes hydration with formation of dextrins and maltose, the latter, at least, being quickly absorbed into the circulating blood. But large quantities of sugar in the blood are certainly inimical to the well-being of the organism, and we find in the liver a tendency for the sugar to undergo a transformation, *i.e.*, a retrogression into glycogen, either through simple dehydration or otherwise. Further, with reference to the possible conversion of proteoses into peptone by the substance of the intestine, we have a perfectly analogous case in the behavior of the intestinal mucous membrane toward maltose, the final product of amylolytic action. Thus, according to the recent work of M. C. Tebb,[1] the mucous membrane of the intestine has the power of transforming maltose into dextrose; simple warming at 40° C. of a solution of maltose in 0.5 per cent. sodium carbonate with a few grammes of the dried mucous membrane from the intestine, being sufficient to insure a marked conversion of maltose into the higher-reducing sugar, dextrose. This observation, I can confirm from experiments just completed in my own laboratory. This action is presumably due to a ferment, which, according to Tebb, is widely distributed throughout the body, being present not only in the intestine, but also in the liver, kidney, spleen, striated muscle-tissue, and, indeed, in the blood-serum; so that it would appear that nearly all the tissues of the body are endowed with the power of transforming maltose into dextrose. These statements being correct, it would seem that, while the amylolytic ferments of the several digestive juices transform, by hydrolytic action, starchy foods into malt-

[1] On the Transformation of Maltose to Dextrose, Journal of Physiol., vol. xv, p. 421.

ose, the latter is exposed during its passage through the intestinal wall, as well as in the blood itself, to another ferment which carries the hydration still further, with formation of dextrose; and yet the latter product is destined, in part at least, to undergo retrogression into a starch-like body, i. e., glycogen, before it is completely utilized by the system. Thus, the analogy between these carbohydrate bodies and the products of proteolysis is complete, and we may well accept the statements already made regarding the ultimate fate of the proteoses and peptones formed during proteolysis, as in no way inconsistent with the general tenor of events going on in the body.

While we are inclined to believe that the chemical changes attending the absorption of proteoses and peptones occur mainly in the epithelial cells of the intestinal mucosa, and that there is a direct transferrence of the alteration-products to the blood, there are still other views that cannot be wholly ignored. Thus, the view originally advanced by Hofmeister,[1] in which special stress is laid upon the functional activity of the leucocytes of the adenoid tissue surrounding the intestine, demands some consideration. The theory supposes that these cells not only have the power of taking up peptones, but also of assimilating and transforming them into the cell-protoplasm. This view being correct, it is plain that the so-absorbed proteid must pass into the circulating blood through the thoracic duct, and Hofmeister further considers it probable that the lymph-cells of the mesenteric glands can transform any absorbed peptone that may escape the leucocytes of the adenoid tissue.

In apparent harmony with this view is the fact that the leucocytes in the adenoid tissue of the intestine are greatly

---

[1] Zeitschr. f. physiol. Chem., Band 4.

increased in number during digestion.[1] Furthermore, it is a well authenticated fact that the proteoses or peptones found in pus are contained in the pus-cells themselves, and not in the fluid in which the corpuscles float.[2] In support of the first statement, Pohl,[3] in his recent study of the absorption and assimilation of food-stuffs, has emphasized the marked increase in the number of white blood-corpuscles in the circulating blood after the ingestion of proteid foods, especially such as meat, Witte's peptone, and gelatin-peptone. It is to be noted further that the increase is most marked at about the third hour after the taking of food, viz., at a time when digestive proteolysis would naturally be at its height. Moreover, the maximal increase, according to Pohl's data, is astounding, amounting as it does in many cases to a hundred per cent. Thus, in one instance, in the case of a dog, the number of white blood-corpuscles per cubic millimetre of blood was 8,689; yet two hours after the feeding of 100 grammes of meat the number increased to 17,296 per cubic millimetre, followed six hours after by a return to the original figure.

This indicates the general tenor of Pohl's results, which have been taken, by some physiologists at least, as confirmatory of Hofmeister's views; the interpretation naturally being that digestive proteolysis in the alimentary tract is accompanied by a rapid production of new leucocytes in the lymph-spaces surrounding the intestine, and followed by a rapid transferrence of the corpuscles from their point of origin to the circulating blood, from which they gradually disappear as their material is made use of in the different parts of the body. In harmony with this view,

[1] Arch. f. Exp. Pathol. u. Pharm., Band 20 and Band 22.
[2] Zeitschr. f. physiol. Chem., Band 4, p. 268.
[3] Arch. f. Exp. Pathol. u. Pharm., Band 25, p. 31.

10

Pohl finds that there is a much larger number of leucocytes in the blood and lymph flowing from the intestine of
an animal in full digestion, than in the arterial blood
coming to the intestinal tract. Further, when due consideration is given to all the circumstances attending the
circulation of the blood through the abdominal organs, in
connection with the great increase in the number of leucocytes during digestive proteolysis, it seems not unreasona-
ble to suppose that some proteid matter might be
transferred from the intestine to the blood during the
digestive period of six or eight hours. Moreover, if
Pohl's views are correct, we see that a portion, at least, of
the proteid food-product may be transformed into organized material in the body of the lymph-cell prior to its
passage into the blood, thus harmonizing with the statement already made regarding the utter lack of proteoses
and peptones in the blood and lymph. This obviously
means an upbuilding of the ordinary products of digestive
proteolysis into the living protoplasm of the leucocytes in
the intestinal walls, implying, however, that the transformation is accomplished solely by the leucocytes themselves, and not by the epithelial cells of the intestine.

I have given this brief summary of Pohl's work because
it is so closely in harmony with the original views of
Hofmeister, and because it offers an easy explanation of
one possible way in which some of the products of digestion might perhaps pass from the intestine into the blood.
I am inclined to believe, however, that the so-called
digestive leucocytosis, which unquestionably does exist, is
not a direct result of digestive proteolysis in general, but
rather an indirect result, coming from the stimulating
action of the nuclein, contained especially in animal cells.
Thus, it is a significant fact, as Pohl himself reports, that
wheat-bread, with its fairly large amount of proteid

matter, and which is fully capable of nourishing the animal body, fails to exert any influence on the number of leucocytes in the blood. Yet we know that the gluten and other proteids of wheat-flour are converted by digestive proteolysis into proteoses and peptones, with the same general properties as like products of animal origin. In this connection we may note the experiments of Horbaczewski,[1] which show that nuclein administered to a healthy man will give rise to a very marked increase in the number of leucocytes in the blood. Thus, a few grammes of nuclein may produce as striking a condition of leucocytosis as a large amount of proteid food, due no doubt to proliferation of the lymphoid elements of all the lymphatic tissues of the body. Horbaczewski has reported that the mere injection of 0.25 gramme of nuclein, in the case of rabbits, will cause marked enlargement of the spleen, with striking karyokinetic changes. Hence, it may be assumed that whenever nuclein is set free in the body, leucocytosis may result, provided the nuclein passes into the circulation and is not decomposed immediately after its liberation.

These facts, it appears to me, offer a more consistent line of explanation of digestive leucocytosis than that advanced by Pohl. All animal foods, especially meat of various kinds and milk, contain considerable nuclein or nucleo-albumins, which, by the action of the gastric juice, are liberated and partially digested, but the nuclein is certainly not dissolved. Nucleins, however, are soluble in weak alkaline fluids, and when exposed to the action of the alkaline pancreatic juice in the intestine, are in great part dissolved. Thus, Popoff[2] has reported that different varieties of nuclein behave somewhat differently in the intestine, according

[1] Monatshefte f. Chemie, Band 12, p. 246.
[2] Ueber die Einwirkung von Eiweissverdauenden Fermenten auf die Nucleinstoffe, Zeitschr. f. physiol. Chem., Band 18, p. 533.

to their origin. In young and tender tissues, solution of the contained nuclein through the alkaline fluids of the intestinal canal is fairly complete, while the older products are somewhat more resistant both to the pancreatic juice and to the putrefactive processes common to the intestine. However, experiments show that the greater portion of the nuclein of ordinary proteid foods is dissolved in the intestine, and absorbed as such in a practically unaltered form. Consequently, passing into the adenoid tissue surrounding the intestine, it has a marked stimulating action on the lymphoid elements, accompanied by a noticeable increase in the number of leucocytes, which are perhaps produced at the expense of a portion of the proteoses and peptones formed during proteolysis.

Thus, my interpretation of these results would lead me simply to the admission that, possibly, a portion of the products of proteolysis might pass from the intestine into the blood-current indirectly, through the bodies of the leucocytes formed in the adenoid tissue of the intestine. But even admitting this, we lack positive proof of any direct transformation of proteoses and peptones into the organized material of the white blood-corpuscles, for it may be that the above products are first transformed through other agencies into serum-albumin, or other like proteids. There are, indeed, many facts which are plainly opposed to any marked absorption and transformation of peptones by the leucocytes of the intestinal mucous membrane. Thus, Heidenhain[1] has severely criticised the theory on the ground that there is very little increase in the flow of lymph from the thoracic duct during absorption, and further that the small percentage of proteid matter in the chyle (about 2.0 per cent.) cannot account

[1] Beiträge zur Histologie und Physiologie des Dünndarms, Pflüger's Archiv f. Physiol, Band 43, Supplement Heft.

for the large amount of proteid absorbed. Further, the objection is made that the leucocytes present in the intestinal mucosa, though numerous, are wholly inadequate to assimilate any large proportion of the ingested proteid food.

Still greater stress, however, may be laid upon the fact that experimental evidence points to the conclusion that lymph-cells cannot assimilate either peptones or proteoses. Thus, quite recently, Shore[1] has studied the results following the introduction of a mixture of such products into the lymphatic system by secretion, by absorption, and by direct injection into a lymphatic vessel. Preliminary experiments on dogs showed that when peptone is introduced into the bile-duct it gradually appears in the lymph of the thoracic duct, consequently this method can be made use of as a means of ascertaining the fate of peptone so absorbed into the lymphatic system. The results obtained, using the ammonium sulphate method for the isolation of the peptone, showed that when peptone is injected into the bile-duct with sufficient force to overcome the low pressure under which bile is secreted, there is an increase in the rate of flow of lymph from the thoracic duct. Further, while peptone is somewhat slow in appearing in the lymph it eventually makes its appearance there, in from sixty to one hundred and forty minutes after its injection into the bile-duct. A certain amount of peptone naturally passes into the blood, but is then rapidly excreted through the urine. When, however, the renal vessels are ligatured, peptone still rapidly disappears from the blood, but then passes into the lymph, and under such circumstances can be detected in the lymph as early as thirty-eight minutes after its injection into the bile-duct. These

[1] On the Fate of Peptone in the Lymphatic System, Journal of Physiol., vol. xi., p. 528.

results, therefore, do not accord with the view that peptones suffer marked transformation by contact with lymph-
cells, for when only three-fourths of a gramme of peptone
is introduced into the bile-duct, unaltered peptone can be
detected in the lymph of the thoracic duct seventy to
ninety minutes after its injection.

With reference to the fate of peptone when it passes by
secretion into the lymphatic system, it will be remembered
that Heidenhain[1] has shown that the injection of peptone
into the blood may be followed by a large increase in the
rate of flow of lymph. Further, the amount of solids in
the lymph, especially of proteids, is considerably increased.
From these and other facts, Heidenhain is led to the view
that the formation of lymph is a true secretion from the
blood-vessels. Shore finds that when small amounts of
peptone are slowly injected into the blood, there is generally only a slight acceleration in the flow of lymph, but
the clotting power of the lymph is affected in a remarkable manner. Thus, about twenty minutes after the commencement of the injection the lymph loses entirely its
power of coagulating. This continues for about twenty
minutes, and then, in spite of the fact that the injection is
being continued, the lymph rapidly regains its power of
clotting, and finally coagulates quicker and firmer than
before. This peculiar action of peptone on the clotting
power of lymph may frequently be observed, even when
the amount of peptone present is too small to be detected
with certainty by chemical methods. Thus, when peptone
in small quantitity is injected very slowly into the blood,
the greater part of it escapes through the urine, but a
small fraction, sometimes too small to actually detect,
passes into the lymph and shows its presence by its peculiar
influence on the clotting of the fluid.

[1] Versuche und Fragen zur Lehre von der Lymphbildung, Pflüger's
Archiv f. Physiol., Band 49, p. 252.

When, on the other hand, peptone is injected rapidly into the blood, 0.3 to 0.6 gramme per kilo. during two to ten minutes, it may disappear completely from the blood in five to ten minutes after the end of the injection. In such a case, the fall of blood-pressure induced leads to more or less arrest of the renal secretion, peptone appearing in large amount in the lymph; but there is no indication of any alteration of the peptone by the lymph, or its contained leucocytes. Thus, when there is no chance for the peptone to escape from the body, as on ligation of the renal vessels, the peptone injected into the blood is rapidly thrown into the lymph, and from the lymph in the tissues of the body it is gradually carried to the thoracic duct, and then again passes into the blood; all of which shows that there is little or no transformation of the peptone by the leucocytes of the lymphatic system.

Further, by direct injection of peptone into a lymphatic vessel, Shore has shown that even so small an amount as 0.049 gramme is not assimilated or transformed by the lymph in half an hour. Consequently, we seem to have strong evidence that peptones are not prone to direct alteration of any kind by the leucocytes of the lymphatic system. Further, it would appear that the lymphoid cells of the spleen are equally unable to assimilate small amounts of peptone injected into the splenic artery. Leucocytes, then, can play no direct part in the absorption of the products of proteolysis from the intestine; the lymph is normally free from both proteoses and peptones, and the leucocytes plainly have no power to transform these bodies into other forms. They can only utilize the proteid material elaborated from the products of proteolysis by other agencies.

Plainly, proteoses and peptones in the blood and lymph are foreign substances. When present in the circulation

they give rise, as we have seen, to an increased flow of .-
lymph and to a change in the coagulability of the blood. ⸰
Further, not only is the flow of lymph augmented but
there is likewise an increase in the amount of solid matter,
while a corresponding decrease is noticed in the solid
matter of the blood-plasma. This fact obviously gives
support to the view that the increased formation of lymph
after the injection of peptone is due to an active process of
secretion by the endothelium cells of the capillary walls.[1]
Further, as we have seen, peptone disappears from the
blood more or less rapidly after its injection, so that it is
quite possible that the loss or alteration of the coagulability of the blood may not be due to the peptone itself,
but rather to an altered condition of the blood induced by
the peptone. Moreover, this altered condition of the blood
may be the real cause of the increased transudation, or
secretion of lymph so conspicuous after the injection of
peptone. Starling, however, by carefully conducted experiments on dogs finds, in conformity with Heidenhain's
views, that peptone injected into the blood exercises a
direct excitatory effect on the endothelial cells, causing
thereby an increased flow of lymph; the increased flow
being in no way caused by the change in the blood that
is simultaneously produced. Further, it would appear,
according to Starling's views, that the change in the coagulability of the blood is not due to the effect of the peptone on the endothelial cells of the blood-vessels, or at
least on their lymph-producing functions. Thus, the
injection of peptone may result in an action on the endothelial cells of the blood-vessels, thereby increasing the
flow of lymph, or on the blood itself with a destruction or
diminution of the clotting power of the blood; the two

---

[1] Starling: Contributions to the Physiology of Lymph Secretion,
Journal of Physiol., vol. xiv, p. 131.

results being more or less independent. Further, the rapid transferrence of peptone from the blood to the lymph is effected by the selective activity of the endothelial cells of the vessel-wall, and according to Starling it is probable that a preponderating part is played by the endothelial cells of the renal capillaries.

In view of all these statements, it is very evident that proteoses and peptones once outside the limits of the alimentary tract may be passed about from organ to organ and from secretion to secretion, inducing changes here and there in their course, but suffering very little change themselves. The main efforts of the system are directed to the removal of these unwelcome strangers as speedily as possible, for their marked physiological action renders them somewhat dangerous visitors.

As normal products of digestive proteolysis, they are never found beyond the limits of the gastro-intestinal canal, but undergo retrogression in their passage through the epithelial cells of the intestinal wall, being presumably converted thereby into serum-albumin,[1] which can be directly utilized for the nutrition of the body; a conversion which is plainly dependent upon certain inherent qualities of the living epithelial cells, and is doubtless of the nature of a dehydration.

[1] See Kronecker and Popoff: Ueber die Bildung von Serumalbumin im Darmkanale, Du Bois-Reymond's Archiv f. Physiol., 1887, p. 345. Also Nadine Popoff, Zeitschr. f. Biol., Band 25, p. 427.

CPSIA information can be obtained
at www.ICGtesting.com
Printed in the USA
BVHW090831220219
540922BV00020B/1102/P

9 781333 341770